Science and Strategies
for Safe Food

Science and Strategies for Safe Food

Surender S. Ghonkrokta

CRC Press
Taylor & Francis Group
Boca Raton London New York

CRC Press is an imprint of the
Taylor & Francis Group, an **informa** business

CRC Press
Taylor & Francis Group
6000 Broken Sound Parkway NW, Suite 300
Boca Raton, FL 33487-2742

First issued in paperback 2020

ISBN 13: 978-0-367-57362-1 (pbk)
ISBN 13: 978-1-138-71327-7 (hbk)

Library of Congress Cataloging-in-Publication Data

Names: Ghonkrokta, Surender S., author.
Title: Science and strategies for safe food / Surender S. Ghonkrokta.
Description: Boca Raton : Taylor & Francis, CRC Press, 2017. | Includes bibliographical references and index.
Identifiers: LCCN 2016056411 | ISBN 9781138713277 (hardback : alk. paper)
Subjects: LCSH: Food adulteration and inspection. | Food contamination--Prevention. | Food industry and trade--Quality control. | Food security.
Classification: LCC TX531 .G47 2017 | DDC 338.4/7664--dc23
LC record available at https://lccn.loc.gov/2016056411

Visit the Taylor & Francis website at
http://www.taylorandfrancis.com

and the CRC Press website at
http://www.crcpress.com

Contents

List of Figures

Foreword

At the outset, I compliment Dr. Ghonkrokta for his venture to bring out an excellent compilation of information on various aspects of food safety in this compact book titled *Science and Strategies for Safe Food*. The subject of food safety is on one hand so commonplace that almost everyone has something to say about it, but on the other hand it is so vast that volumes can be written on each of its topics. When it comes to policy making, or paying attention to it in our daily lives, no subject gets such fleeting treatment as food safety, despite it being so crucial to our well-being and even that of future generations (due to genotoxic contaminants). At the international level, concerns have been expressed about access to sufficient and nutritious food by millions in developing countries. It has been recognized that unsafe food is no food. Food security has been defined to include "safe and nutritious food" that meets the dietary needs and food preferences of the population for an active and healthy life. The World Food Summit held in Rome in 1996 set a target of "food security for all" in "an ongoing effort to eradicate hunger in all countries, with an immediate view to reducing the number of undernourished people to half their present level no later than 2015." This goal was reiterated by the Millennium Summit in South Africa in September 2000, the largest gathering of world leaders in history, and the World Summit on Food Security held in Rome in 2009. Significantly, in 1996, it was estimated that about 800 million people, particularly in developing countries, did not have enough food to meet their basic nutritional needs. In 2009, the Declaration of the World Summit on Food Security noted: "We are alarmed that the number of people suffering from hunger and poverty now exceeds 1 billion. This is an unacceptable blight on the lives, livelihoods and dignity of one-sixth of the world's population." The Declaration committed itself to supporting national, regional, and international programs that contribute to improved food safety and animal and plant health, including prevention and control of transboundary animal and plant pests and diseases. It also committed to promoting effective national food safety systems, covering all stages of the food chain and involving all actors, that ensure the compliance of food products with science-based international standards, and that improve food safety and quality for the people, with emphasis on locally available food. The Codex Alimentarius Commission has also prepared the Strategic Plan 2014–2019 to meet the challenges of emerging food safety issues arising from increased globalization, climate change, shifting population, and so on. It is against this background that I find Dr. Ghonkrokta's book to be a timely, relevant, and positive contribution in the direction of assuring safe food for all.

To successfully achieve the global objective of alleviating hunger, a robust regulatory mechanism is necessary, though not sufficient, particularly in developing countries where the hiatus between the reach of the regulator and the large number of actors in the food supply chain is not likely to be bridged soon. This is not to suggest that regulation is the ideal, capable, or desirable mode of ensuring the safety of food available to the consumer at the end of the chain. Experience the world over has shown that voluntary compliance and awareness among food business operators as well as consumers can be the only sustainable strategy to achieve the goal of food safety. India adopted this approach in the year 2006 when the Food Safety and Standards Act (FSSA) was enacted by the Indian Parliament and as a result of which the Food Safety and Standards Authority of India (FSSAI) was established in the year 2008. The old Prevention of Adulteration Act (PFA), 1954, provided for a minimum punishment of three months' imprisonment for any food-related offense, including an offense like the sale of substandard and misbranded food. In 2011, the FSSA promulgated more persuasive and voluntary compliance by food business operators and the decriminalization of minor deviations from the regulations that do not make food unsafe. As director of the FSSAI, Dr. Ghonkrokta was a part of the team that ushered India into the new FSSA regime. Being an academic from the field of food technology and a member of the civil service, and having worked in the FSSAI at the crucial juncture of transition from the PFA to the new Act, he is in a unique position to appreciate food safety issues from the perspectives of food science, enforcement, and food business operators. In this book, he has skillfully combined all these perspectives and discussed the tools required to achieve the global mandate for food safety. He has defined the role of actors in the entire food chain and takes us through the entire gamut of issues pertaining to the production and consumption of food by us. The book is a handy reference for regulators and food safety managers in different sectors, from primary producers to processing, transport, retail, and distribution, as well as the food services sector.

The most remarkable thing I find about this book is that Dr. Ghonkrokta has been successful in explaining the concepts of food safety in a most lucid manner, distilling his academic knowledge and administrative experience so that it can be understood by practitioners in the field and laymen with equal ease. I am sure that the book will be of immense use for all stakeholders in food safety and will fill a long-felt gap in the integrated and holistic treatment of the subject of food safety in a single volume.

I once again congratulate Dr. Ghonkrokta for bringing out this informative and refreshing book, which is written in a comprehensive, simple, and readable style, and which will go a long way to promoting, inculcating, and establishing a culture of food safety.

V.N. Gaur
Member, Central Administrative Tribunal, New Delhi
Former CEO, Food Safety & Standards Authority of India

Preface

Through IT revolutions, globalization and information-sharing systems have done a lot of good for creating awareness of and demand for healthy food and health consciousness. The consumer of the twenty-first century wants to know all about the benefits and health issues related to the food he/she is eating. No doubt the taste and smell of food are relished and enjoyed, but the missing link of associating food with health is still gaining currency. Now consumers want to know about the composition of food, the ingredients it contains, and the health benefits or harm it can do to the body so that informed choices can be made. The response to all these demands and aspirations of consumers has been very positive. The concerns of a calorie-conscious generation have resulted in modifications to food handling, upgrading of food regulations, requirements of systems for quality maintenance, more labeling information, declarations of ingredients, and better processing techniques. The growing realization of the desire to interlink food security, environment, health, and sustainable development has brought food safety to the forefront.

I and most of my professional fraternity have understood and treated food safety as a very important aspect of the food science and technology profession. We have always been very careful about the quality and safety of the product—more so because of the requirements of the job, legislation, and also to maintain brand value—and the underlying reason for both has appeared to be economic, even when all of us have genuinely wanted to be part of a system that was striving to provide consumers with safe and wholesome food. It was beyond our imagination that even a street vendor or a small restaurant would be organizing and managing food safety issues more professionally.

It was a great eye-opener when, during my road journey between Delhi and Chandigarh, the owner of a *Dhaba* (a small, inexpensive restaurant) displayed his knowledge of and appreciation for safe food. He informed me that he always buys fresh produce from nearby village farms so that his food tastes good and is also safe. He takes personal interest and care that the fruits and vegetables are sourced from known farmers who use manure instead of fertilizers. He explained that the vegetables grown with fertilizers have a different taste. He also knew which of the farmers in the area were using fewer pesticides. He may not have been aware of good agricultural practice (GAP) certification, but what he was following and narrating was exactly the process as per the requirements of GAP—he was practicing GAP without certification. He also informed me that whatever he cooked daily was either consumed the same day or reprocessed/reheated to give to his dairyman, so he was not wasting the food but utilizing it as cattle feed, and also ensuring that it was not contaminated so as not to infect the cattle. He also knew a lot about the grains, cereals, spices, and other products that he had to buy from the market as they were not available fresh. His was the best example of self-regulation. I could make out that he was proud of his profession, loved to do it in the best possible way, and was eager to learn. His ultimate words were, "Sir, even if we provide food to anyone, it is of no use if it is not safe and nutritious." I was touched by these simple words, which have a very deep meaning. Food, which is essential for survival, loses its significance if it instead turns out to be a cause of disease and misery. While traveling in a developing country like India, we always try to avoid taking foods from simple roadside restaurants, fearing that the owner may not be following basic hygiene practices, and thus we also avoid food products that are not boiled and served hot, like tea. This may be true most of the time, but there are examples like the one I just narrated of people who have the passion and involvement to supply safe food. Such good practices, which are practical, simple, and inexpensive, need to be replicated. I am sure there must be a large number of people who are following best practices without any expectations and only for their personal satisfaction toward the job. Sometimes, in our race to follow certain set-down procedures, we miss out on the capability and ingenuity of simple, normal-looking people who are contributing to the society in a big way. Sometimes these people need to be appreciated and proliferated.

Another such example came from discussions with some of my colleagues on washing hands before eating food and how to inculcate these habits in children. The issue drifted to the use of antibacterial hand sanitizers. One of the scientists opined that it is not appropriate to use these sanitizers, that their regular use should be avoided, and that they must only be used in exceptional circumstances. The reason for this is that use of these antibacterial gels, which can get ingested with food, can depress the body's immune system. Secondly, but no less seriously, these antibacterial products are now thought to contribute to the development of antibiotic-resistant strains of bacteria, which are becoming a serious threat to public health. These doubts appeared to be genuine, so the safety we are looking for in hand sanitizers may be misplaced.

These two incidents, which happened in two different settings, with two different individuals having different views on issues, had one common concern: food safety. Both had their own perceptions and understanding of the issues as per their education, experience, and the requirements of their profession. They each found their own solutions and strategies to address their concerns. Both of them were passionate about their jobs, had concern for the safety and well-being of people, and were serving society in their own way. One thing that struck me was that there is a need to share experiences and learn from each other's best practices and innovations; this way we can collectively identify the problem, find out the scientific explanations/reasons responsible, look for simpler solutions, and offer rationales for the actions/solutions so that long-term strategies can be prepared.

These incidents, and many more similar ones, were my motivating factors to pen something on the topic. During my association at the newly created Food Safety & Standard Authority of India as Director of Enforcement, during interactions with food business operators, professionals in the food processing field, and teachers and students of Food Science and Technology, it emerged that they were all looking for information in a comprehensive form that could help them plan and make strategies for producing safe food. Many students who wanted to take up jobs in the regulatory field or as food auditors, some teachers who wanted to start courses on food safety, and other friends from the industry who wanted employees with certain skill sets expressed their desire to have one book that could address issues concerning food safety and quality in a concise form. Food business operators are interested in having a successful food safety program to protect their brand name and ensure a steady growth in return on investment. Measures taken to ensure food safety should help the food industry to stay in compliance with the relevant local, state, and global food codes. These food safety measures should make sense economically and should also help in maintaining lower operational costs in the long term. Some enlightened and inquisitive consumers, especially those representing consumer organizations, had many questions regarding the safety of food available in the market. Everyone was interested in having a comprehensive and integrated system for achieving food safety. We all know that there is a need to establish a robust food safety system in any organization, starting from selecting a food safety team to developing systems, training, execution, verification, and gap analysis. The implementation and acceptance of changes become easier if the principles, rationales, and logic behind such changes are understood. My background in regulation, academia, and industry has helped me immensely to understand these concepts and principles and to present them in simpler forms. Driven by my desire to promote and protect public health, taking into consideration the realities of a food business and the concerns of the consumers, I have made my attempt to present available research and knowledge so that food safety can be adopted as a culture into every setting. As a food scientist, if we want to see a positive change in universal food safety, all professionals must find ways to make behavioral changes at the consumer as well as the food business operator level. If consumers become more demanding, the industry is bound to deliver. Food scientists and technologists have to act as the catalysts and major driving force behind bringing in the desired change.

Dr. Surendar S. Ghonkrokta, IAS (R)
New Delhi, India

Author

Dr. Surender S. Ghonkrokta, IAS (R), has work exposure of more than 35 years in the food industry, university, and civil services. Notable among these are his appointments as director, Food Safety & Standards Authority of India (FSSAI); secretary, Home & Secretary Industries, Govt. of Arunachal Pradesh; and special commissioner (Food & Supplies), Govt. of Delhi. In the FSSAI, he was responsible for the implementation and enforcement of the Food Safety and Standards Act 2006, coordination of State Food Safety Commissioners, accreditation of laboratories and Food Safety Management System (FSMS) agencies, capacity building in food safety, and policy initiations for a food safety plan and food control system in India. He had a stint at Himachal Pradesh Horticulture Produce & Marketing Corporation's (HPMC) World Bank and assisted with the country's first apple concentrate plant at Parwanoo as a quality control specialist. Subsequently, he joined the University of Horticulture and Forestry at Solan, H.P., as assistant professor and remained in teaching until he joined the civil services in 1985. As a civil servant, he served in various capacities in the areas of revenue, taxation, education, criminal justice, environment, election, transport, and tourism.

Dr. Ghonkrokta is former president of the Association of Food Science and Technology (India). He has a PhD, MSc (Food Technology) from Central Food Technological Research Institute (CFTRI), Mysore, MBA (FMS, Delhi University), and MA (Rural Social Development, University of Reading, UK). Dr. Ghonkrokta has published 20 papers in national/international journals, authored one book, and contributed chapters to two books.

He headed the Indian delegation for the Codex Committee Meeting on Food Additives (CCFA) at Hangzhou, China, and the Codex Committee Meeting on Nutrition and Foods for Special Dietary Uses (CCNFSDU) at Bad Soden am Taunus, Germany. He has represented India on food safety/nutrition issues and attended more than 200 workshops, seminars, and conferences within and outside India.

Dr. Ghonkrokta has training associations with the Emergency Management Institute of the Federal Emergency Management Agency (FEMA), USA; the Department of Homeland Security, USA; the Asia Pacific School of Economics and Government, Australian National University (ANU), Canberra, Australia; the Indian Institute of Management (IIM), Ahmedabad; the National Institute of Financial Management (NIFM), Faridabad; the Institute of Applied Manpower Research, New Delhi; the Indian Institute of Tourism Management (IITM); and the Indian Institute of Criminology & Forensic Sciences, New Delhi. He is a life member of the Association of Food Scientists & Technologists of India (AFST), the Indian Institute of Public Administration (IIPA), and the Association of British Scholars, India. He was adjunct professor/visiting faculty at the Department of Food Sciences, Central University of Pondicherry, and GGS IP University, New Delhi. He is visiting faculty at various national and international institutes within India.

1 Food, Environment, and Health

Food is any substance, whether simple, mixed, or compounded, that is ingested via eating or drinking or used as confectionery or condiment. Food is essential for all living things. Our physiological needs define the basic determinants of food. Humans need energy in order to survive, which leads to feelings of hunger and, later, satiety. Specific signals—for example, blood depleted of nutrients or an empty stomach—trigger hunger and create the need for eating food. As food is eaten, the body responds, resulting in a state of no hunger, called satiety. The balance between hunger/stimulating appetite and food intake/satisfying appetite/bringing satiety is controlled by the central nervous system. Satiety signals play a role in influencing the timing of the next meal, and can also influence the size of that meal. Satiety is also known to play a role in energy regulation. Food is consumed to provide necessary energy to maintain bodily functions and is also essential for growth. The type of food and nutrient requirements may vary across and within species. Food is usually of plant or animal origin and constitutes roughage, water, carbohydrates, fats, proteins, vitamins, and minerals. Human beings have specific requirements, and these components of food are required in fixed quantities. Historically, human beings secured food through two methods: hunting and gathering and foraging from forests. Subsequently, humans learnt to domesticate animals and started practicing agriculture. Today, the majority of the food energy required by more than seven billion people of the world is supplied through agriculture. Even in the case of animal sources of food, there is an indirect contribution from plants as animals feed on plants or food derived from plants. All over the world, cereal grain in the form of rice, wheat, millet, or maize constitutes a majority of daily sustenance than any other type of crop. Corn (maize), wheat, and rice account for 87% of the world's grain production. All grain is not directly consumed as food. Most of the grain that is produced worldwide is fed to livestock, so the process of producing animal-origin food is longer and more complex. Foods that are not from animal or plant sources include edible fungi, especially mushrooms, and ambient bacteria used in the preparation of fermented and pickled foods like leavened bread, alcoholic drinks, cheese, pickles, and yogurt. There are a few blue-green algae such as spirulina that are also used as food. Inorganic substances and chemicals such as salt, baking soda, and cream of tartar are used to preserve or chemically alter an ingredient and thus indirectly become part of food. There are a large number of additives, preservatives, and aids that are included in food for various reasons other than nutrition. Certain other substances are taken in with food unintentionally, such as adulterants. Most legal interpretations define food as a substance, whether processed, semiprocessed, or unprocessed, that is intended for human consumption. Going by this logic, water can be called food too. However, according to some sources, such as the Food Safety & Standards Act, 2006 of India, under section 3 (j), it has been clarified that the water that is not in package form, bottled, or not being used during food processing will not be considered as food. Similarly, medicine is not food as it is not taken to satisfy hunger and the human body does not rely on medicine for energy or growth.

1.1 COMPONENTS OF FOOD

Food has different components, and each component has a specific role to play in various bodily functions. The major components of our food are

1. Carbohydrates
2. Fats
3. Proteins
4. Vitamins
5. Minerals
6. Water
7. Roughage

Carbohydrates, fats, and proteins are the major nutrients. They are consumed in larger quantities as they are the major sources of energy. Vitamins and minerals are required in smaller quantities. A deficiency of these nutrients can cause a variety of diseases. Though water is an important constituent of our food and makes up for two-thirds of our body

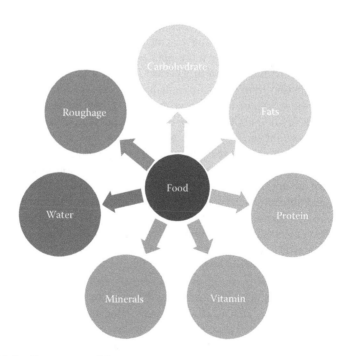

FIGURE 1.1 Components of food.

weight, it is usually not considered a nutrient. Roughage does not have a direct role to play, but it helps in digestion. Our diet usually contains the entire range of these nutrients in varying amounts. Figure 1.1 gives a picture of the components of food.

1.1.1 CARBOHYDRATES

Carbohydrates are compounds made up of three elements: carbon, hydrogen, and oxygen, the proportion of hydrogen and oxygen being the same as in water, that is, 2:1. Carbohydrates, which are mostly in the shape of glucose, sucrose, and starch, are the primary source of energy. When these are oxidized in the body, carbohydrates produce energy. These are generally divided into two categories: simple carbohydrates, which are digested quickly, and complex carbohydrates, which are digested slowly. Dietary fiber is another form of carbohydrate required for proper digestion.

The carbohydrates in our food are obtained mainly from plant sources like wheat, rice, maize, potatoes, peas, beans, and fruits. The world's three main cereal crops that provide us with carbohydrates are wheat, rice, and maize, all of which are abundant in starch. Sources of simple carbohydrates include fruits, sugars, and processed grains such as white rice or flour. Milk contains a sugar called lactose, and sugar is also obtained from other foods. We can find complex carbohydrates in green or starchy vegetables, whole grains, beans, and lentils.

1.1.2 FATS

Chemically, fats are esters of long-chain fatty acids and an alcohol called glycerol. Fats are made of three elements: carbon, hydrogen, and oxygen. Fats differ from carbohydrates in constitution as they contain a smaller proportion of oxygen; fat consists of three molecules of a fatty acid and one molecule of glycerol. Fats are members of a heterogeneous group of organic compounds known as lipids. The main function of fats in the body is to provide a steady source of energy, and fats provide twice the energy provided by the same amount of carbohydrates. Fats have a special quality as they can also be stored in the body as fat reserves for subsequent use. The fats present in our food cannot be absorbed by our bodies as such because they are complex organic molecules that are insoluble in water.

Despite the belief that fats are considered bad for health, they are required and essential for general health. Fats help our bodies to synthesize fat-soluble vitamins such as vitamin D. Healthy fats include monounsaturated and polyunsaturated fats. Nuts, olives, and avocados are sources of monounsaturated fats. Fish and seafood are primary sources of polyunsaturated fats. In addition, vegetable oils such as canola contain both monounsaturated and polyunsaturated fats. Certain types of fats are bad for our health, such as trans-fat and saturated fat, both of which increase the risk of cardiovascular diseases. It is always advised that, as far as possible, the intake of saturated fat and trans-fat should be minimized if it cannot be excluded completely.

Foods like butter, oils, milk, and eggs are major sources of fat. The food that we eat in our diet contains dietary fats in different proportions. Some of the foods may have less than 1% fat; some are rich sources of fat. Different food products have different fatty acids. The major fatty acid present in butter is butyric acid. The major fatty acid present in coconut oil is lauric acid, which is also a saturated fatty acid. The major fatty acid present in animal fat is stearic acid. The major fatty acid present in plant fats is oleic acid. Therefore, certain food products are rich in some fatty acids and may have different proportions of saturated and unsaturated fats.

1.1.3 Proteins

Proteins are highly complex organic compounds made up of carbon, hydrogen, oxygen, and nitrogen. Some proteins also contain elements such as sulfur and phosphorus. Proteins are required to build and repair our bodies. Thus, these are vital for the growth of children and adolescents. Proteins are needed to repair and maintain the wear and tear of body tissues in adults. In addition to all this, proteins also supply energy to the body. Proteins are made up of nitrogen-containing compounds called amino acids. Amino acids link through peptide bonds to form protein molecules. There are more than 20 of these amino acids and they all occur in almost all proteins, but the relative amount of each amino acid differs in different proteins. It should be noted that the proteins consumed through food are not used by our bodies in their original form. This is because of two reasons: firstly, proteins are insoluble in water, and secondly, they are very complex molecules. When food is digested in the small intestine, the proteins present in the food are broken down into simpler substances called amino acids. Amino acids are water-soluble and less-complex molecules. The amino acids thus formed are absorbed from the intestine into the blood. The blood carries these free amino acids to various body cells, where they are regrouped to form specific proteins such as skin, muscle, blood, and bone.

As proteins are required for the growth and repair of the body, they are essential for various biochemical reactions and activities in the body. The function of enzyme proteins is to catalyze the biochemical reactions taking place in the body, like digestion. The main enzyme proteins in the body are pepsin and trypsin. The function of hormone proteins, such as insulin, is to regulate the various body functions. The function of transport proteins, such as hemoglobin, is to carry different substances from the blood to various tissues in the body. Contractile proteins, such as myosin and actin, help in the contraction of muscles and other cells. Structural proteins, such as collagen, form the structural elements of the cells and tissues of our bodies. Protective proteins, such as gamma globulins present in the blood, help fight infection in our bodies. The properties or functions of food proteins depend on the amino acids of which they are made. There are different combinations of amino acids in any particular protein. Some proteins contain all the amino acids required by our bodies, whereas others contain only some of them. Some of the amino acids are called essential amino acids as these are crucial for the body. Protein is required for healthy muscles, skin, and hair. In addition, protein contributes to normal chemical reactions within our bodies. We can get proteins from plant sources as well as animal sources. Some sources of plant proteins are groundnuts, beans, whole cereals like wheat and maize, and pulses. Some of the best sources of animal proteins are lean meat (i.e., meat without fats), fish, eggs, milk, and cheese. These are all body-building foods. The most valuable proteins are found in milk and eggs. These are valuable because they contain all the essential amino acids required by our bodies. These proteins are particularly needed by children because children are growing and proteins are the main building blocks of the body. Complete sources of protein, primarily meats, contain the nine amino acids essential for human health. Plant-origin foods such as rice and beans also provide our bodies with the nine essential amino acids.

1.1.4 VITAMINS

Vitamins are complex organic compounds essential for health and thus considered primary components of nutrition. Vitamins do not provide energy to our bodies but act as catalysts in certain metabolic chemical reactions, which lead to normal growth and good health. Vitamins are necessary for normal growth, good health, good vision, proper digestion, healthy teeth, gums, and bones, and for life to be maintained. Though vitamins are only needed by our bodies in minute quantities, their presence is essential in our diet. When vitamins were discovered, their chemical compositions were not known immediately, so, initially, the vitamins were represented by letters like A, B, C, D, and so on. More than 15 vitamins are known at present and each one of these is needed for a specific purpose in the body. Important vitamins include vitamins A, B Complex, C, D, E, K, and folate.

Many fruits and vegetables have high vitamin content, as do fortified dairy and bread products. Most vitamins cannot be made by the body, so they have to be supplied through various foods that contain them. Only two vitamins—vitamin D and vitamin K—can be prepared in the body; all the other vitamins are prepared in plants. Almost all food items contain more than one vitamin in varying amounts. With advances in science, all the vitamins can be produced synthetically.

Some vitamins are soluble in water, whereas others are soluble only in fats or oils, so, on the basis of their solubility, all the vitamins can be divided into two classes or groups: water-soluble vitamins and fat-soluble vitamins. Vitamins B Complex and C are water-soluble vitamins. Vitamins A, D, and K are fat-soluble vitamins. The chemical name of vitamin C is ascorbic acid. It is a water-soluble vitamin. Vitamin C is necessary for keeping teeth, gums, and joints healthy. It also increases the body's resistance to infection and helps fight diseases. The various sources of vitamin C are amla, limes, oranges, and tomatoes. The chemical name of vitamin D is calciferol. Vitamin D is a fat-soluble vitamin. Vitamin D is necessary for normal growth of bones and teeth because it increases the absorption of calcium and phosphorus in the body. The various sources of vitamin D are milk, fish, eggs, and butter. Vitamin D is also produced in the body when the skin is exposed to sunlight. The chemical name of vitamin E is tocopherol. Vitamin E is necessary for moral reproduction, normal functioning of muscles, and protection of the liver. The various sources of vitamin E are green leafy vegetables, milk, butter, tomatoes, and wheat-germ oil. Vitamin K is a fat-soluble vitamin, which is known as phylloquinone. Vitamin K is necessary for normal clotting of blood and for preventing hemorrhage. The various sources of vitamin K are green leafy vegetables like spinach and cabbage, tomatoes, and soybeans. Consuming too much of vitamin K can also result in serious toxicity. This is also true of a few other vitamins such as vitamin B-6 or vitamin A.

1.1.5 MINERALS

Minerals are the essential nutrients that the body needs but cannot produce; they include the inorganic substances found in foods. Minerals are needed to build bones and teeth, form red blood corpuscles, coagulate the blood, and for the functioning of muscles, nerves, thyroid glands, and so on. Several minerals are needed as catalysts for enzymes to do their work. They are essential for the metabolic activities of the body. Our bodies can use minerals in the compound form and not as pure elements. For example, we cannot take sodium metal or chlorine gas in their element forms because they are toxic (poisonous) and can even kill a person, but as a compound, sodium chloride, for example, is a mineral salt that is harmless and, in fact, essential for our bodies. Most of the mineral intake of humans is sourced from various plants, as plants take the various minerals from the soil through their roots and supply these to humans and animals through the food chain, so even the minerals that we get from some animals are, in fact, derived from the plants that the animals eat. Some of the important minerals needed by our bodies are: iron, iodine, calcium, phosphorus, sodium, potassium, zinc, copper, magnesium, chlorine, fluorine, and sulfur. Iron is the most important mineral required by our bodies. Iron is needed to prepare a protein called hemoglobin, present in blood. This hemoglobin helps us in transporting oxygen from the lungs to the body cells through the blood. Some major sources of iron are liver, kidneys, bajra (pearl millet), ragi (finger millet), and eggs. Iodine is another important mineral needed by our bodies. Iodine is needed in small quantities for the preparation of a thyroid hormone called thyroxin. Some of the major sources of iodine are fish, seafood, and iodized salt. Calcium salts are required for making bones and teeth, helping with blood clotting, and for the proper working of the muscles. The major sources of calcium are milk, milk products like cheese, beans, green leafy vegetables, whole gram, meat, fish, and ragi (finger millet). Phosphorus is also required for the formation of bones and teeth, and for the conversion of carbohydrates in energy. Phosphorus is important because it is a compound of ATP, DNA, and RNA. The major sources of phosphorus are milk, vegetables, bajra (pearl millet), ragi (finger millet), and nuts.

Deficiencies of minerals can result in serious health conditions such as brittle bones and poor blood oxygenation. Like vitamins, overdosing on minerals can result in life-threatening conditions—for example, a potassium overdose can cause improper kidney function.

1.1.6 WATER

The human body is composed of 60% water, and our brains are composed of 70% water. Water is necessary to maintain proper bodily function. An overdose of water is also not good and in severe cases can be fatal.

1.1.7 ROUGHAGE

Roughage, also known as dietary fiber, is cellulose and is a type of carbohydrate that forms the cell walls in plants. However, it is not a food as such, because cellulose cannot be digested or absorbed by the body. When eaten, however, cellulose acts as roughage and helps in keeping the intestinal tract in good working order. Cellulose helps to maintain a healthy digestive system. Dietary fiber foods keep one feeling full for hours after a meal. Legumes, whole grains, and berries are good sources of dietary fiber.

It is important to note here that most of the food that we consume is not used by our body cells in its original form. As discussed earlier, this is because most of the ingredients of our bodies are complex molecules that are insoluble in water, and hence cannot be absorbed by the blood. The main function of the digestive process is to convert the food that we eat into forms that can be easily absorbed and assimilated by our bodies. Our food contains mainly carbohydrates, fats, and proteins—all complex organic compounds, which are insoluble in water and hence cannot be absorbed by the blood directly and assimilated by the body. During the digestion process, carbohydrates, fats, and proteins undergo mechanical and chemical treatment, due to which they are spilt into simpler forms. These simpler molecules are absorbed by the intestinal canal and are ultimately transported through the blood to the places where they are required for various activities.

1.2   FOOD SECURITY: HUNGER AND ACCESS TO FOOD

Hunger and food insecurity pose large and complex problems, in part because they are closely tied to poverty, a condition that has prevailed since the beginning of recorded history. Hunger has been defined as the discomfort, weakness, illness, or pain caused by the lack of food. At the 1974 World Food Conference, the term "food security" was defined with an emphasis on supply. According to the 1974 World Food Conference, food security is the "availability at all times of adequate world food supplies of basic foodstuffs to sustain a steady expansion of food consumption and to offset fluctuations in production and prices." Later definitions added demand and access issues to the definition. The World Food Summit of 1996 defined food security as existing "when all people at all times have access to sufficient, safe, nutritious food to maintain a healthy and active life" (FAO, 2008). Commonly, the concept of food security is defined as including both physical and economic access to food that meets people's dietary needs as well as their food preferences. Food security is a complex sustainable development issue, linked to health through malnutrition but also to sustainable economic development, environment, and trade. The situation is such that the world now has more food-insecure and nutrient-deficient people than it did a decade ago. With the degradation of nutritional quality, the prevalence of lifestyle diseases like obesity-related diabetes, high blood pressure, and cardiovascular diseases is increasing at a very rapid rate. Expanded food production has done little to address the fact that, in developing countries, between one-third and one-half of all deaths among children under five are still related to malnutrition. Poor food supply management is a major problem, as we have enough food but it does not reach the needy. According to the United Nations Food and Agriculture Organization (FAO), 20%–30% of food produced globally is lost every year. That's enough to feed an additional 3–3.5 billion people. There is wastage of food in a long food chain at every stage. This also results in the wastage of essential resources like water and soil, which are used to produce the food that goes to waste and is not used by anyone.

A study by the FAO has pointed out that, although there is no doubt that significant progress has been made in meeting the Millennium Development Goals on poverty and hunger, almost a billion people still live in extreme poverty (living on less than $1.25 per person per day) and 795 million still suffer from chronic hunger. If we are to eradicate poverty and address hunger, there is an urgent need to do much more to achieve the new Sustainable Development Goals on eradicating poverty and hunger by 2030. Most of the extreme poor live in rural areas of developing countries and depend on agriculture for their livelihoods. The population in some parts of the world is so poor and malnourished that families live in a cycle of poverty that passes from generation to generation. Many developing countries are adopting a successful new strategy for breaking the cycle of rural poverty, combining social protection and agricultural development. Social protection measures—such as cash benefits for widows and orphans, and guaranteed public works employment for the poor—can protect vulnerable people from the worst deprivation. These measures basically aim to increase households' purchasing power and can also allow them to increase and diversify their diets. Greater income can help them to save and invest in their own farms or start new businesses. Agricultural development programs that support small family farms in accessing markets and managing risks can create employment opportunities that make these families more self-reliant and resilient. Social protection, agricultural development, and working together can break the cycle of rural poverty. Social protection programs, regardless of type, can effectively reduce food insecurity. Such social protection programs ultimately help raise consumption levels and result in greater dietary diversity at the household level. Studies have shown that women are often the main beneficiaries of social protection programs and play a key role in household food security and nutrition. It has been claimed that programs that target women, consider their time constraints, and enhance their control over income have stronger food security and nutrition impacts, especially for children (FAO, 2015).

The United Nations (UN) recognized the right to food in the Declaration of Human Rights in 1948, and has since noted that it is vital for the enjoyment of all other rights. The World Summit on Food Security was held in 1996; the FAO called the summit in response to widespread undernutrition and growing concern about the capacity of agriculture to meet future food needs. The conference produced two key documents: the Rome Declaration on World Food Security and the World Food Summit Plan of Action. The UN Committee on Economic, Social, and Cultural Rights (1999) affirmed that the right to adequate food is indivisibly linked to the inherent dignity of the human person and is indispensable for the fulfillment of other human rights enshrined in the International Bill of Human Rights. It is also inseparable from social justice, requiring the adoption of appropriate economic, environmental, and social policies at both the national and international levels, oriented to the eradication of poverty and the fulfillment of all human rights for all (UN, 1999). Another World Summit on Food Security took place in Rome between November 16 and 18, 2009. The rights to food and food security were becoming issues that countries were not able to handle; the number of poor and malnourished persons was increasing (FAO, 2009). In 2009, the World Summit on Food Security

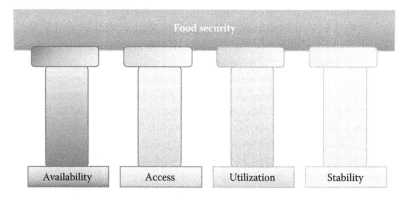

FIGURE 1.2 The pillars of food security.

stated that the "four pillars of food security are availability, access, utilization, and stability." Figure 1.2 depicts the four pillars of food security.

1. *Availability*: Food availability relates to the supply of food through production, distribution, and exchange. As per the data available, per capita world food supplies are more than adequate to provide food security to everyone on Earth, yet still food is not available to some persons at the same cost, all year round, and in adequate quantities, as it is to others. This makes food accessibility a huge barrier in achieving food security. Production of food is dependent on a variety of factors like land use, soil management, crop selection, and techniques in production and management. The productivity of crops can also be affected by changes in rainfall and temperature. The use of land, water, and energy to grow food often competes with other uses, including growing crops for non-food uses, which can affect food production. Land used for agriculture can be used for urbanization or lost to desertification, salinization, and soil erosion due to unsustainable agricultural practices. Different types of food crops are grown in different parts of the world and food must be distributed to different regions or nations. Food distribution involves the storage, processing, transportation, packaging, and marketing of food. There is spoilage and wastage in the food chain from farm to fork. Food chain infrastructure, storage technologies, transport arrangements, and retail systems affect the amount of food wasted in the distribution process. These processes can make food expensive and costly, and thus unavailable.

2. *Access*: Food access has wide connotations as it includes affordability, the allocation of food, and the food preferences by the consumers. The cause of hunger and malnutrition is not always a scarcity of food but an inability to access available food, usually due to the inadequacy of purchasing power. Adequate resources are required to purchase food at prevailing prices,

when a person does not have other resources like land or water with which to produce food for themselves. Location and many other factors can also affect access to food. Therefore availability is not enough alone; the food should also be accessible to people.

3. *Utilization*: The quantity and quality of food that reaches members of the household is a very important factor for food security. In order to achieve food security, the food ingested must be safe and there must be enough to meet the physiological requirements of each individual. Utilization of food is dependent on the preparation, processing, and cooking of food. The nutritional values of the household determine food choice and psychological and social well-being.

4. *Stability*: Food stability refers to the ability to obtain food over time. This can be long-term or short-term depending on the circumstances. Food may be unavailable during certain periods of time. At the food-production level, floods, droughts, and other disasters result in crop failures and decreased food availability. Instability in markets resulting in rises in food prices can cause food insecurity. Other factors that can temporarily cause food insecurity are loss of employment or productivity, which can be caused by illness. Therefore the stability and consistency of availability and access to food is essential.

Numerous efforts and initiatives have been undertaken to correct the shortcomings of supply, increasing understanding that merely making more food available will not assure better food security, nutrition, or health at the household and individual levels. Biofuel production, such as Jatropha in Africa, now competes with food for land, and climate change is already negatively impacting crop yields, with major implications for food supply. For low-income consumers all over the world, high and more volatile food prices, such as those seen in 2007, are also resulting in food insecurity. Poor consumers respond by purchasing cheaper food that is low in nutrients and thus causing more distress to their health and life.

Unfortunately, the concept of food security has mostly focused on food calories, and the emphasis has been only to achieve the minimum number of calories per individual. However, the quality and nutritional value of food is equally important. The concept of food security is evolving over time and now also includes nutritional status in terms of protein, energy, vitamins, and minerals for all household members at all times. This transformation has brought back the focus on food safety and nutritious food.

1.3 FOOD SYSTEMS, NUTRITION, AND PUBLIC HEALTH

Good nutrition is very important for good health and disease prevention. Food is essential for everyone but has special significance for the growth and development of children and adolescents. In the case of grown-up persons, evidence suggests that a diet of nutritious foods could help to reduce incidences of heart disease, cancer, and diabetes, leading causes of death even in developed countries. Knowledge of the benefits of a balanced diet persuades some people to change their eating habits and lifestyles. For others, eating a healthy diet may not be easy because healthy food options are not readily available, easily accessible, or affordable. Organic foods are becoming popularized on the premises that these are more safe and healthy, as they are free from chemical fertilizers and pesticides. The prices of these foods are prohibitive for middle- and lower-income classes. The purchasing capacity of poor people is also a major issue. There is enough food for everyone, but the food should also be safe and wholesome; it is not enough merely to have food that can provide adequate calories. A diet is required to be balanced and must contain vitamins and minerals as well. Water is a very essential part of our diet. Most of the water required by the body is supplied as part of food, as most foods contain 70%–90% water. Other than this, water is taken separately. Use of contaminated water has been a cause of concern all over the world. Unsafe water has been responsible for a large number of illnesses and deaths, especially in developing countries. The water used for processing food has to be safe as well; otherwise the food prepared with contaminated water will not remain safe. Sanitation and hygiene are also very important and must be followed at every stage of food production. The process of taking adequate precautions starts the moment the process of cultivation of the crop begins. Subsequently, one has to track the process at every stage of the cultivation, production, and harvesting of the crop to the time it is finally eaten. The path to better health and nutrition must look beyond the availability of food at affordable prices, clean water, and good sanitation, and consider behavioral factors such as time constraints for women in low-income households. In fact, malnutrition and foodborne diarrhea have become a double burden as they take up most of the attention and resources of any country, especially developing nations, which face problems in these two areas.

Access to sufficient amounts of safe and nutritious food is key to sustaining life and promoting good health. Unsafe food containing harmful bacteria, viruses, parasites, or chemical substances causes more than 200 diseases ranging from diarrhea to cancer.

Foodborne and waterborne diarrheal diseases kill an estimated 2.2 million people annually, most of whom are children. Diarrhea is the most common, acute symptom of foodborne illness, but other serious consequences include kidney and liver failure, brain and neural disorders, reactive arthritis, and cancer (WHO, 2014). According to the Millennium Development Goals Programme Report 2015, "while some countries have made impressive gains in achieving health related targets, other countries are falling behind, mainly those affected by high levels of HIV/AIDS, economic hardship or conflicts." In other countries, progress has been limited because of conflict, poor governance, economic or humanitarian crises, and lack of resources. The effects of the global food, energy, financial, and economic crises on health are still unfolding, and action is needed to protect the health spending of governments and donors alike. Undernutrition is an underlying cause of about one-third of all child deaths. Over the past year, rising food prices coupled with falling incomes have increased the risk of malnutrition, especially among children. Although the percentage of underweight children under 5 years of age declined globally from 25% in 1990 to 18% in 2005, subsequent progress has been uneven. In some countries, the prevalence of undernutrition has increased, and in 2005 stunted growth still affected about 186 million children under 5 years of age worldwide. Globally, child mortality continues to fall. In 2008, the total annual number of deaths in children under 5 years old fell to 8.8 million—down by 30% from the 12.4 million estimated in 1990. In 2008, mortality in children under 5 years old was estimated at 65 per 1000 live births, which is a 27% reduction from 90 per 1000 live births in 1990 (WHO, 2010). The data from the Foodborne Disease Outbreak Surveillance System (USDA, 2013)shows the human impact of foodborne disease outbreaks in the United States. In 2013, 818 foodborne disease outbreaks were reported, resulting in 13,360 illnesses, 1062 hospitalizations, 16 deaths, and 14 food recalls. Outbreaks caused by *Salmonella* increased 39% from 2012 (113) to 2013 (157). Outbreak-associated hospitalizations caused by *Salmonella* increased 38% from 2012 (454) to 2013 (628). The most common single food categories implicated in all foodborne outbreaks were fish (50 outbreaks), mollusks (23), chicken (21), and dairy (21, with 17 due to unpasteurized dairy products). The most common causes of outbreaks were viruses (35% of reported outbreaks) and *Salmonella* (34%) (CDC Surveillance Report, 2013). If this is the situation in the United States, which has an established food control system, things are surely much worse in countries that have virtually no food safety apparatuses in place. There are other diseases that are also attributed to diet imbalance, food safety, and food security. These are nutritional deficiencies and noncommunicable diseases like cardiovascular diseases, kidney ailments, liver disorders, and cancers. Nutritional deficiency diseases are caused by deficiencies in minerals and vitamins. This explains why food systems, nutrition, and public health have a close, overlapping correlation and are dependent on each other. Figure 1.3 explains the relationship between food systems, nutrition, and public health.

Nutritional interventions are necessary to correct imbalances in the diet. Diet diversity is incredibly important for good nutrition. Agricultural researchers and food production companies need to look at a number of different commodities, not just the major food staples. There are many methodologies that can be adopted to provide wholesome and complete food. In many countries, health problems related to dietary excess due to imbalanced food are an ever-increasing threat. Families influence children's dietary choices and risk of obesity in a number of ways, and children develop food preferences at home that can last well into adulthood (Gruber and Haldeman, 2009). Just as employed adults spend most of their day at work, children spend much of their day at school. In the United States, the National School Lunch Program and related federal school meal programs, administered by the U.S. Department of Agriculture (USDA), serve more than 30 million children every day, including breakfast, lunch, and after-school snacks (USDA, 2013).

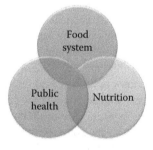

FIGURE 1.3 Relationship of food systems, nutrition, and public health.

Researchers have found that participating in the School Breakfast Program is associated with lower body mass index (BMI) in children, while participating in the lunch program does not affect obesity. Students participating in the School Breakfast Program are also less likely to skip breakfast, which may reduce the risk of them becoming overweight by spreading food intake more evenly across the day (Gleason et al., 2009). Close proximity of fast-food restaurants to schools has been linked to increased risk of obesity in schoolchildren (Davis and Carpenter, 2009).

If we take the example of India, which has the largest number of poor people in the world and a large number of malnourished children, is slowly experiencing the change, as is the case with rest of the developing nations. But the pace of change is very slow. India, which is developing quickly, has resources and technological capabilities but is a nation of contradictions. The rich have access to the best food and health facilities, but the poor exist with all the deprivations. India is experiencing epidemiological and demographic changes mainly driven by sustained economic development, which has boosted incomes and reduced poverty. Over the past 50 years, male life expectancy has increased by more than 30 years. In the past decade, infant mortality has decreased by 20% nationwide, with some states achieving declines of 40%. Despite this, the rate of increase in the incidences of disease is increasing, especially the so-called lifestyle diseases. There has been an increase in the rates of chronic disease burdens, such as cardiovascular diseases, cancer, and tobacco-related illnesses. In the case of other diseases, too, the population continues to suffer health problems associated with poverty and ignorance, like water- and foodborne diseases, vaccine-preventable diseases, pregnancy and childbirth-related complications, and malnutrition. There have been wide disparities in health gains between the prosperous and impoverished states and areas within the country as well. Improvements in health parameters in India have been lower compared to neighboring South Asian countries. For example, the infant mortality rate in India was 65 per 1000 live births in 2002, having declined 42% since 1980; it was lower still in Bangladesh, Indonesia, and Nepal at 48, 32, and 62, respectively, having declined 63%, 60%, and 50% over the same period (WHO, 2010). One of the major reasons for slow improvements in health indicators can be attributed to food safety and food security issues. In 2005, India launched the National Rural Health Mission to provide accessible, affordable, and quality healthcare to its rural areas, particularly to poorer and more vulnerable populations. The central goal of the mission is to expand public health services, improve infrastructure and staffing, and reduce the burden of health spending on the country's poor. The mission aims to help bridge the wide gaps in health between affluent and poorer states, to sustain the health gains in the better performing states, and to address the chronic disease burden that will increasingly strain India's healthcare system. Target interventions should also address disease conditions that are the major sources of infant and childhood mortality and infectious disease burdens, to better address the needs of those who are underserved by the current system. In 2006, the Ministry of Health & Family Welfare, India, came out with the Food Safety & Standards Act, which was an improvement on the existing regulations already being enforced. The new Act was responding to the serious concerns expressed by consumers about the quality and safety of the food available (FAO, 2015; Government of India, 2011).

1.4 IMPACT OF DIET AND FOOD CHOICES ON OUR ENVIRONMENT

Food choices and diet have a direct impact on the environment. These can be classified in the following way.

1. *Toxicity and pollution*: "Pollution" is a word often used to describe the problems and challenges of the environment in many parts of the world. The livestock sector is a major player, responsible for 18% of greenhouse gas emissions measured in carbon dioxide equivalent. This is a higher share than transportation. The livestock sector accounts for 9% of anthropogenic carbon dioxide emissions. The largest share of this derives from land-use changes, especially deforestation caused by expansion of pastures and arable land for feed crops. Livestock are responsible for much larger shares of some gases with far higher potential to warm the atmosphere. The sector emits 375 Kt of anthropogenic methane (which has 23 times the global warming potential of carbon dioxide); most of that comes from enteric fermentation by ruminants. The sector also emits 65% of anthropogenic nitrous oxide (with 296 times the global warming potential of carbon dioxide), the great majority of which comes from manure. Livestock are also responsible for almost two-thirds (64%) of anthropogenic ammonia emissions, which contribute significantly to acid rain and the acidification of the ecosystem. This high level of emissions opens up large opportunities for climate change mitigation through livestock actions (FAO, 2006). Other than this, the fertilizers and pesticides that are used for producing crops are major sources of pollution and the destruction of plant and animal life, especially aquatic life. When we talk of the food environment, the word "toxin" immediately comes to mind because we are aware of the chemical contaminations and pesticides prevalent all over. A U.S. National Academy of Science report designates nitrogen and phosphorous pollution as the main threat to U.S. coastal waters. Many of the fresh water streams are now unsuitable for aquatic life primarily due to excess nutrients. The basic reason of this pollution is that fertilizer runoff from corn production used primarily to feed livestock and leaching from large manure ponds. This runoff of excess nutrients makes it difficult for the fish to live.
2. *Overexploitation of resources*: Over the past several decades, the business of commercial fishing has been transformed from familiar, family-run operations to industrial fishing operations employing immense factory ships that are owned by large corporate holding companies with little concern for conservation. The result has been a 90% decline in the abundance of valuable top-predator fish species. Moreover, 82% of all major world fisheries are now either fully exploited, overexploited, or in recovery, according to a 2010 FAO report (FAO, 2010). Indiscriminate use and wastage of water in irrigation and other uses results in a shortage of water. It is estimated that by 2025 about 64% of the aquatic population will be living in water-stressed basins. Land is also becoming less available as the population is

increasing and a greater area is being put under cultivation. Other than that, land is used for urbanization, growing nonfood crops, and other uses as well. Forests are also depleting, which in turn affects the water table.

3. *Loss of biodiversity and complementary crops*: We are going through an unprecedented biodiversity crisis. The rate of disappearance of living species is between 50 and 500 times in various biodiversity zones. Due to demand, single-crop growing is being adopted. This affects the supply of nutrients, land degradation, and scarcity of water.

Today the situation is such that, especially in cities, air and water are becoming the major sources of the spread of pollution. Even though the food itself is usually safe to consume, the world in which most consumers live makes choosing healthy food very hard and choosing unhealthy food very easy. It's truly a toxic environment that eats away at healthy lifestyles. There has been a lot of awareness among consumers about eating safe food. Consumers are focusing on planting and buying fresh fruits and vegetables. There is a demand for subsidizing the cost of healthy choices. The food environment plays a major role in the food choices people make, even for the most independent-minded consumer. There has been a lot of awareness and discussion of so-called junk foods and the presence of pesticides in fruits and vegetables, or antibiotics in honey or milk. All these debates and discussions have resulted in positive changes. Any positive changes to the food environment can begin to shift momentum away from a world that so easily promotes unhealthy eating, and toward a world where healthy eating is the default choice. Health, nutrition, and food security cannot be achieved without a focus on the synergistic linkages and interactions between individuals, societies, and their environments. Social, economic, and physical environments have an influence on the overall food environment. The economic environment operates within the broader society and includes food marketing, economic and price structures, food production and distribution systems, transportation, and agricultural practices and policies. Collectively these environments influence what food choices we make and where and how much we eat. Food choices are intertwined with and dependent on the community and environment.

There are issues that have a direct impact on the environment. What we eat and how we grow it is very important. Whatever we eat has a cost for the environment, as the type of food and its quantity will impact the soil and water, and food crops will directly affect these resources. How we grow the food will determine how many pesticides and other toxins are added to the food and soil. It is natural that most of the population, especially in developing countries, will have a larger demand for cheap food, so the food policies of most of the developing countries focus on providing large amounts of inexpensive calories. In India, two of the cheapest sources of calories are wheat and rice, which the government has subsidized and which make up a large percentage of calorific intake today. Wheat and rice can also be efficiently grown by farmers in India, but growing just one crop consistently (i.e., a monoculture) depletes the soil and forces farmers to use greater amounts of pesticides and fertilizers. The effects of pesticides and fertilizers on natural wildlife and water supplies are well documented. Pesticides and herbicides are environmental toxins known as xenobiotics. Xenobiotics include not only pesticides and herbicides but also plastics (e.g., bisphenol A), surfactants used in food packaging, household chemicals, industrial chemicals (e.g., PCBs and dioxins), and heavy metals (e.g., lead, mercury, and cadmium). These products have been shown to have a negative impact on health. Increasingly, the food we eat comes from far away. Now we are using a lot of food that is processed. Processing of food contributes to air and water pollution and also depletes the food of its nutrients. Ecologically, when crops are grown and taken away we are depleting the soil of these nutrients as we are not replacing them fully. Commercial fertilizers only add nitrogen, phosphorus, and potassium but provide little else, and ultimately this changes the biochemistry of the soil. We have broken the ecological link wherein the nutrients from the soil used in growing food are consumed locally and then returned to that soil as compost and other waste. The USDA has been tracking the nutritional quality of produce since the 1950s and has seen that there has been a steady decline in nutrition quality over the years. According to Brian Halweil (2007), a researcher for World Watch, vitamin C has declined by 20%, iron by 15%, riboflavin by 38%, and calcium by 16%, so we are now getting less nutrition per calorie in our food. In essence, we have to eat more food to get the same vitamin and mineral content. Now farmers are caught in a cycle of fertilizers and pesticides; crops are now dependent on these two. Genetic modifications, changes in varieties, use of more fertilizers, and also changing soil conditions have made plants more vulnerable to pests. To meet the increased demand for food due to the increase in population, farmers have had to enhance productivity and thus use more pesticides, which introduce those chemicals into our air and water supplies and our bodies.

1.5   WORKING TOGETHER TOWARD A FOOD SYSTEM THAT NURTURES HEALTH AND SUSTAINABILITY

We have a food system that appears to have a lesser concern for health and the environment, and a health system that has a lesser concern about food. There is a need to change this paradox and bring some cohesiveness. There is a need to bring necessary changes to the food system that can take care of nutrition and healthy food in a sustainable manner. Food security and nutrition are two faces of the same coin and the cornerstones of sustainable development. They are necessary for living, learning, prospering, staying healthy, and contributing to the well-being of the self and the development of the society. They are also essential for fulfilling the aspirations for innovation, inclusive economic growth, and human development. Hunger, malnutrition, and lack of access to clean water are both symptoms and causes of poverty. All these deprivations contribute to insecurity and instability among the poor. Ending poverty and hunger is a prerequisite to sustainable and sustained economic growth, reducing preventable deaths and improving health. Such a situation will result in improvements in education, empowerment to the deprived, and environmental well-being. The major challenge has been to ensure appropriate and comprehensive actions are taken that can tackle all of the multiple dimensions of food security and nutrition. People who have no access to food are sufferers of hunger and are deprived of food. Unfortunately, in many developing countries, these deprived people include those who play a major role in food production, such as farmers. Although the growth of food production has slowed down in recent years, we have been repeatedly reminded that globally there is still enough to feed everyone. We need to ensure that the world's poor have the means to obtain the food they need at a reasonable cost, throughout the year, especially in times of crisis. Supportive governments, emboldened political will, and an understanding of society all have to unite to recognize the fundamental right to food. The cooperation and coordination of all in society is required to end hunger and ensure healthy and productive populations. The challenge is to ensure that everyone has access to food all year round, and no children under 2 years of age are stunted. Today sustainability is a major global challenge and will be an even greater challenge as the population of the world grows. The FAO has estimated that, to meet the requirements of the increased population, agriculture output will have to be increased by 60% by 2050. This increase in production is to be achieved with probable risks of climate variability and change. To address the issue of hunger and nutrition, there will be a need to change production and consumption practices. By doing our best to diversify what we eat, we not only better serve our bodies but also the environment, soil, and oceans. Food security and good nutrition can be sustainable if our way of life is sustainable. We have to shift to more sustainable production approaches, which can ensure higher yields with a lesser impact on the environment. Food producers produce and supply as per the demand of the consumer. Today consumers demand a wide selection of produce throughout the year, even when local growing seasons don't support its availability. In such situations food must be imported from faraway places, which increases the carbon footprint of our meals. This puts pressure on natural resources and distorts the natural balance.

Agriculture and food systems have to be more sustainable and resilient. These changes can be brought about by efficient management of water, land, and other natural resources. Global warming and other challenges concerning the environment are to be addressed. Farmers in developing countries who are suffering due to flooding or drought have to be supported. Special attention and support has to be extended to such rural displaced populations linked to agriculture. Efforts must be taken to halt and reverse land degradation and desertification. Changes in strategy are required to manage water resources sustainably and protect the forested, aquatic, and mountainous ecosystems, increase energy efficiency, and reduce the carbon footprint of agriculture. Investment in agriculture has to be increased so that the rural population is able to innovate and adopt improved agricultural practices. Small farmers are agents of transformation, so they need to be strengthened. Many countries are innovating and taking steps to improve the environmental impacts of our food system, and environmental organizations are encouraging sustainable farming practices and working toward lowering the system's carbon footprint. At the same time, physicians are helping patients understand the connection between diet and health, and other groups are raising awareness of the social responsibility we bear for our food choices. Data shows that in developing countries 70% of malnourished people are in rural areas. They do not produce enough food and also do not have enough money to buy food. Better governance and appropriate support should be provided to transform the economically depressed rural class. The poor should have enough security to enable them to innovate, add value, and increase investment. The Healthy and Sustainable Food Program aims to advance improvements in our food system by providing multiple industries with a common framework for creating a positive impact on human communities and the environment. Coordinated efforts are required to break the cycle of poverty and malnutrition. Increased investments in agriculture, food, and nutrition will translate into an improved environment, increased incomes, and prosperity.

We will have two billion more people to feed by 2050. Land, water, and other resources for food production are already under stress. There will be a need to take a few drastic steps. The work to look for alternate sources of food like insects and other similar creatures has been started. Indigenous, traditional, and local foods that were discontinued over a period of time are being rediscovered to be used as food items. These crops maintained balance in the environment and biodiversity. People may have to shift to vegetarian diets because more resources like land, water, and nutrients are required to produce the same amount of energy or protein from animal sources as compared to vegetarian food. In addition to changing food choices, we have to adopt a multipronged strategy by addressing the issues of food safety, food security, and nutrition. These issues are inextricably linked and ultimately have an impact on the health and well-being of the people. Unsafe food creates a vicious cycle of disease and malnutrition, particularly affecting infants, young children, the elderly, and the sick. Foodborne diseases impede socioeconomic development by straining healthcare systems and harming national economies, tourism, and trade. Food supply chains now cross multiple national borders. Only collaboration between governments, producers, and consumers will be able to ensure food safety and healthy population.

REFERENCES

CDC Surveillance Report. (2013). Annual summaries of foodborne outbreaks. http://www.cdc.gov/features/foodborne-diseases-data/, Accessed on August 12, 2015.

Davis B. and Carpenter C. (2009). Proximity of fast-food restaurants to schools and adolescent obesity. *American Journal of Public Health* 99: 505–510.

FAO. (2006). *Livestock's Long Shadow: Environmental Issues and Options*. Rome, Italy: Food and Agriculture Organization of the United Nations.

FAO. (2008). An introduction to the basic concepts of food security. www.fao.org/docrep/013/al936e/al936e00.pdf, Accessed on March 6, 2017.

FAO. (2009). *Declaration of the World Food Summit on Food Security*. Rome, Italy: Food and Agriculture Organization of the United Nations.

FAO. (2010). *State of World Fisheries and Aquaculture*. Rome, Italy: Food and Agriculture Organization of the United Nations.

FAO. (2015). The State of Food and Agriculture 2015 (SOFA): Social Protection and Agriculture: Breaking the Cycle of Rural Poverty. Rome, Italy: Food and Agriculture Organization of the United Nations.

Gleason P. et al. (2009). *School Meal Program Participation and Its Association with Dietary Patterns and Childhood Obesity*. Princeton, NJ: Mathematica Policy Research, Inc.

Government of India. (December 2011). Annual report to the people on health. Government of India, Ministry of Health and Family Welfare. http://mohfw.nic.in/WriteReadData/1892s/6960144509Annual%20Report%20to%20the%20People%20on%20Health.pdf, Accessed on January 31, 2016.

Gruber K.J. and Haldeman L.A. (2009). Using the family to combat childhood and adult obesity. *Preventing Chronic Disease* 6: A106.

Halweil B. (2007). *Still No Free Lunch: Nutrient Content of U.S. Food Suffers at Hands of High Yields*. Foster, RI: Organic Center.

UN. (1999). *The Right to Adequate Food*. Geneva, Switzerland: United Nations Committee on Economic, Social, and Cultural Rights. www.un.org/documents/ecosoc/docs/2001/e2001-22.pdf, Accessed on October 5, 2015.

USDA. (2013). *National School Lunch Program Fact Sheet*. Washington, D.C.: United States Department of Agriculture. www.fns.usda.gov, Accessed on August 2015.

WHO. (2010). *World Health Statistics 2010*. Geneva, Switzerland: WHO Library Cataloguing-in-Publication Data.

WHO. (November 2014). Food safety-fact sheet. http://www.who.int/mediacentre/factsheets/fs399/en/. Accessed on October 24, 2015.

2 Principles of Food Safety

2.1 FARMING, SUPPLY CHAIN, AND FOOD SAFETY

Food products such as bread, milk, meat, fruits, vegetables, sugar, and so on originate from farms. These are either produced directly on farms as fresh produce or processed further to give them various shapes/names, but all are agricultural produce. Farmers grow food crops that they harvest, store, and transport to markets or to processing plants for preservation and transformation into a variety of food products. Farming entails the use of natural resources, land, and water to produce food for human and animal consumption. This basic concept of farming has remained unchanged over time; yet agriculture has been witnessing changes since man learnt to grow crops and started rearing animals. In this century alone, new technologies and methods have been developed that have greatly increased the variety of agricultural inputs available to farmers and also enhanced the productivity of farming. As a result, today's rapidly growing global population is being fed by a constantly declining population of farmers. The pace of change taking place in agriculture has been different for different regions in the world. Whether in developed economies or underdeveloped nations, there has been increased reliance on the use of machinery to improve operating efficiency, and on natural and synthetic chemicals both to fight pests and diseases and enhance crop growth. These technologies not only affect the food farmers grow but also impact the environment. Consequently, farmers have multiple and complex responsibilities: to supply primary agricultural products with acceptable levels of man-made chemicals in large enough quantities at affordable prices, and to minimize any damage caused to the environment. Agricultural processes are becoming more complex due to increased demand for food for the growing population. There has been an increase in the use of chemicals, as farmers need fertilizers to supplement natural soil nutrients, antibiotics to prevent animal diseases, and pesticides to protect crops against animals, insects, weeds, and various microorganisms. All these chemicals, which are passed on to the food produced on the farms, contribute to making this food unsafe. The primary objective of food production was to produce sufficient quantities of food at affordable prices, but as awareness about health issues and productivity in agriculture has increased, the consumer has begun to focus more attention on the quality and safety of the food and the way in which it is handled and produced.

Farming practices also vary between countries, regions, and even individual farms within the same region. Farmers can use different cultivation techniques to have either a single or mixed type of farming. One can have an exclusive dairy farm, piggery, or goat farm. Similarly, one can have an apple farm, or any fruit and vegetable farm. Someone can practice pisciculture or beekeeping separately or in combination. In developing nations where most of the population of the country is engaged in and dependent upon

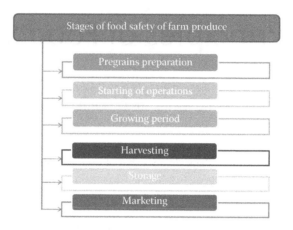

FIGURE 2.1 Food safety of farm produce.

agriculture, combined agricultural activities are normally undertaken. This is done mainly to diversify and spread the financial risk, and this combining of activities further complicates the process of growing crops. But whatever the activities of the farmer, some agricultural processes are basic and remain the same, although maybe under different names or in different forms. These stages/activities (Figure 2.1) are as follows:

1. *Pregrowing preparations*: Preparing and maintaining the land (or water or pasture) for growing the desired crops. This preparatory phase is very important as it lays the foundation for safe food. The type of seed or animal is also critical.
2. *Starting of operations*: The sowing of seeds or the multiplication of animals or fishes. What kind of feed is used for the animals, what treatment is given to the seeds, or which variety of fruit crop is to be planted are very important.
3. *Growing period*: Protection of the crop is very important. It is in this phase that maximum chemicals are used to keep plants free from pests and to achieve good growth/production. In the case of animal husbandry, this stage involves protecting the animals' safety and health to ensure they will gain weight, give birth, produce milk, or lay eggs.
4. *Harvesting*: Whatever has been grown is harvested. Here, the parts that are to be used for food are segregated, mainly by mechanical processes. Milking animals, collecting eggs, and harvesting beehives are a few processes followed for specific foods.
5. *Storage*: Since harvesting is done to a quantity that cannot be used in one go, the food must be stored before it is used. Storing crops, housing animals, and transporting both to protect them against pests and disease also involve the use of chemicals and other processes that are not environmentally friendly.
6. *Marketing of produce*: The selling of crops and animals directly for human consumption or for use as raw materials or ingredients in the food processing industry is decided according to requirement.

All these six stages through which food is moved from farm to consumer are important from the food safety point of view. The safety, integrity, and quality of the produce is to be maintained at every level. Figure 2.1 explains the sequence of six stages through which the food safety of farm produce is ensured.

Food safety is the overall quality of food that makes it fit for consumption. The quality and safety of the final food product is dependent on the care and treatment that it gets at each of the previously mentioned stages of production, beginning with the genetic potential of the plants grown or the animals raised. This, however, is just the beginning, since farmers must provide an environment in which plants and animals can develop to their full genetic potential. This involves matching animals and plants to the natural environment, providing plants with nutrients, and animals with the food necessary to ensure optimal growth. The genetic potential of an apple, for instance, is responsible for its size, color, taste, smell, and nutritional value. To achieve this potential, the farmer cares for the apple tree constantly. He supplements natural soil nutrients when needed and protects the tree and the fruit against insects and disease. As a result of this care and attention on the part of the farmer, the consumer is supplied with fruits that are insect- and fungi-free, and whose appearance and taste meet the consumer's expectations.

2.2   AGRICULTURE PRACTICES AND FOOD SAFETY

A consumer normally chooses or rejects a food product based on its taste, smell, appearance, and consistency. The consumer is not able to judge the wholesomeness and safety of the food, as these cannot be assessed through sensory evaluations. Many food items, especially so-called junk foods, are not so healthy, but people still eat them due to how they look and taste. Sometimes even food that is contaminated cannot be judged as such due to the masking of the contamination effect through spices or other ingredients that can mask the bad odor of spoilage. Spoilage, which can be a sign of unsafe food, may be noted, but even here appearances can be deceptive. In general, there are two situations in which food could be called safe: when there is an absence of chemicals at levels that may be harmful to health and when there is an absence of microorganisms and their toxins in amounts that may cause illness. Chemicals may be toxins that naturally occur in the farm product itself, or they may come from external sources as a result of farming practices, like pesticides or chemicals in the soil and atmosphere. Microorganisms include bacteria, parasites, and viruses. The priority for farmers is to ensure that their products—fruits, vegetables, or animal-origin foods—are produced in a safe manner. To comply with this, they may receive advice from experts on the correct use of fertilizers, pesticides, antibiotics, and other products whenever these are used for crop growing and animal husbandry. Chemicals that are used in agriculture and animal husbandry—such as pesticides, herbicides, insecticides, antibiotics, and growth hormones—are subject to strict regulation. They undergo rigid testing procedures before they are accepted for use in farm practices. The safety of these chemicals is assessed, and the quantity for use is also fixed so that they are not overused. This testing must prove that the chemical is

- *Effective*: The chemical should be able to achieve the intended objective in the prescribed quantity. A product should benefit the plant or animal it is intended to help, with no negative effects on other species.
- *Safe for human beings*: The negative side effects in humans, either during use on the farm or from residues that may remain in food, must be within safe limits.
- *Minimally damaging to the environment*: The chemical should have minimum impact on the surrounding environment and should not leave any harmful residues in the soil or water.

Diseases can also be transmitted through foods. Certain microorganisms can contaminate crops and remain dormant as residues in food and ultimately cause diseases in human beings. Similarly, there are a few microorganisms that can cause harm to animals and, when the meat is eaten by human beings, may cause disease. Milk is boiled or pasteurized to protect humans against the possibility of contracting tuberculosis (TB) from cows. During slaughter, animals are inspected in an effort to prevent the TB bacterium from reaching the consumer. Sometimes animals carry microorganisms that can cause diseases in humans but do not manifest at any stage in the life of the animal. *Salmonella, Listeria, E. coli*, and Campylobacter are examples of such organisms that can harm the human body without manifesting in the host animal. Animals may acquire these microorganisms from their feed or water, through insects, birds, or rodents, and from other sources. To limit the spread of microorganisms to their animals, farmers observe strict hygienic practices on the farms during slaughter and subsequent transportation and packaging. In the case of other food products as well, processing methods, cooking, pasteurization, sterilization, appropriate storage, and packaging are adopted in such a way that the chances of the presence of a disease-causing microorganism is ruled out, especially if there is any doubt about a product's safety.

Because the consumer expects food to be readily available and reasonably priced, as well as wholesome and safe, agricultural research is an ongoing process that focuses on fulfilling consumer expectations. Research activities include developing new plant varieties to provide greater nutritional value, higher production efficiency, or resistance to pests and disease. These and other objectives of agricultural research aim at further improving the quality and safety of food as it leaves the farm, to minimize any negative impacts on the ecology and to contribute to the goal of sustainable agriculture.

2.3 COMMON HAZARDS TO THE SAFETY OF FOOD

There are a few terms that are relevant when discussing food safety. "Microorganisms" are organisms of microscopic or submicroscopic size that are the main sources of the contamination or biological spoilage of food; these will be discussed in detail. "Food infection" is a microbial infection resulting from the ingestion of contaminated food. "Food intoxication" is a type of illness caused by toxins. Under certain conditions, some bacteria and fungi can produce chemical compounds called "toxins." Toxicity in food can also come from chemical hazards, mainly through pesticides. "Food spoilage" occurs when the original nutritional value, texture, or flavor of the food is damaged and the food becomes harmful to people and unsuitable to eat. This can happen because of three different categories of hazard:

1. *Physical*: These include contaminants such as pieces of glass or metal, which have the potential to cause physical injury if consumed. Once the contaminant is present in the foodstuff, the risk can only be eliminated if the contaminant is noticed and removed. A few foreign objects that accidentally find their way into food are hair, wire, and dust. To prevent physical contaminants from entering the food chain, one should wear hair restraints and avoid wearing jewelry while preparing, cooking, and handling food. Physical hazards are divided into two basic categories:
   a. Intrinsic physical hazards that are found naturally in the food (e.g., bones in fish or meat).
   b. Hazards from technical faults during cultivation, harvesting, transportation, and/or processing (e.g., foreign bodies like metal, glass, and plastic in raw materials, or nuts and bolts from engineering problems, faulty packaging, and so on).
2. *Chemical*: These include contaminants such as toxic chemicals, unsafe levels of cleaning and sanitizing chemicals, or toxins that may be produced following contamination by certain organisms such as bacteria or fungi. In some cases, killing or deactivating the organism may not destroy an existing toxin, a chemical substance that can cause foodborne illness. Toxicity may enter the food chain through use of metals, pesticides, cleaning products, sanitizers, preservatives, and other additives. To prevent chemical contaminants from entering the food chain, it is required to teach employees how to use chemicals and store them in their original containers to prevent accidental misuse and leakage into food, and to make sure that labels clearly identify the chemical contents of containers. Care has to be taken that hands are washed thoroughly after working with chemicals. Foods, especially fruits and vegetables that are eaten raw, must be washed properly in cold running water. Pest control should be operated properly to ensure that chemicals do not contaminate foods. Utensils and equipment, which can contain potentially toxic metals like lead, copper, brass, zinc, antimony, and cadmium, should be carefully used as highly acidic foods such as tomatoes or lemons can react with metals. Highly acidic foods such as tomatoes or lemons can react with metals. One of the main sources of entry of harmful metals into food is through the use of contaminated water, which may contain pesticides and heavy metals from the environment. Chemical hazards can be divided into three categories:
   a. Intrinsic chemicals that are found naturally in food (e.g., aflatoxins, mycotoxins, hemagglutinins in red kidney beans, and poisons in mushrooms or Japanese fugu fish)
   b. Added materials that might cause hazards for the consumers due to overdose or reaction with other materials (e.g., sodium nitrite, coloring agents, animal drugs, preservatives, pesticides, and fungicides)

c. Chemicals from technical faults during cultivation, harvesting, transportation, and/or processing (e.g., poisonous material from packaging,
cleaning agents, metals dissolved in the product, maintenance materials, lubricants, paints, and coatings)

3. *Biological*: Most of the spoilage in food and food products is caused by
biological agents. These include organisms such as bacteria, viruses, fungi,
and parasites.

a. *Bacteria*: Bacteria are single-celled living microorganisms responsible for many plant and animal diseases. Bacteria reproduce through
"binary fission," when one cell divides to form two new cells. All
bacteria exist in a vegetative stage. Some bacteria have the ability to
form a spore in which they can survive in adverse or extreme conditions; these are called "spore-forming bacteria." Bacteria are "pho-
tosynthetic," meaning they have the ability to make their own food
through the use of sunlight; thus bacteria also give off oxygen. An
average bacterium measures 1 μm. There are different kinds of bacteria. Spoilage bacteria break down food so it looks, tastes, and smells
bad, thus it is undesirable and unacceptable to eat. Pathogenic bacteria
are disease-causing bacteria that can make people ill if they or their
toxins are consumed with food. It is essential to understand the phases
of the growth of bacteria so that we are aware of the process of growth
and can calculate the rate of food spoilage. The first phase is called
the "lag phase," as during this phase bacteria adapt themselves to the
growth conditions. This is the period in which the individual bacteria
are maturing and are not yet able to divide. The second phase is called
the "log phase," "logarithmic phase," or "exponential phase," as this
is the phase in which growth is very rapid and the bacteria double in
number every few minutes. The third phase is the "stationary phase,"
in which the growth rate slows down as a result of nutrient depletion
and the accumulation of toxic products. This phase is reached as the
bacteria begin to exhaust the resources that are available to them. The
fourth and last phase is the "death phase" or "decline phase." During
this phase the bacteria run out of nutrients and die. To spoil food,
bacteria need to grow and multiply. Bacteria feed on proteins and
carbohydrates; foods that contain these items can support the growth
of microorganisms. "Potentially hazardous foods" (PHFs) have the
potential for contamination as they have the characteristics to allow
microorganisms to grow and multiply.

b. *Viruses*: Viruses are the smallest of the microbial food contaminants,
and they rely on a living host to reproduce. Viruses can't metabolize
nutrients, produce and excrete wastes, move around on their own, or
even reproduce unless they are inside another organism's cells. Viruses
are the simplest and tiniest of microbes; they can be as much as 10,000
times smaller than bacteria. Viruses come in many sizes and shapes, and
consist of a small collection of genetic material (DNA or RNA) encased
in a protective protein coat called a "capsid." The disease caused by

viruses differ from bacteria. Viruses can only multiply inside the living host and do not multiply in foods. Viruses are usually transferred from one food to another and from a food handler to food and water. A PHF is not needed to support the survival of a virus. Hepatovirus or the Hepatitis A virus is found in the human intestinal and urinary tracts and contaminated water.

c. *Fungi*: Fungi are a group of organisms and microorganisms that are classified within their own kingdom—the fungal kingdom—as they are neither plant nor animal. Fungi draw their nutrition from decaying organic matter, living plants, and even animals. Many play an important role in the natural cycle as decomposers and they return nutrients to the soil; they are not all destructive. Fungi can be single-celled or multi-celled microorganisms that can cause food spoilage. Generally, these contaminants will only cause health problems if they are alive and able to infect the consumer when ingested. Fungi usually reproduce without sex. Single-celled yeasts reproduce asexually by budding. Examples of fungi are as follows:

   i. Molds, which cause spoilage in food and can cause illnesses. They grow under almost any conditions but grow well in sweet, acidic foods with low water activity. Freezing temperatures prevent or reduce the growth of molds but do not destroy them. Some molds produce toxins called "aflatoxins."

   ii. Yeasts, which also cause food spoilage. Yeast spoilage produces a smell or taste of alcohol. Yeast is also used for the fermentation of food.

d. *Parasites*: Parasites are organisms that need a living host to survive. In fact, many bacteria, viruses, and fungi also qualify as parasites, but here we are dealing with parasites other than these. A parasite is an organism that lives by feeding upon another organism. Parasites living in the human body feed on our cells, our energy, our blood, and the food we eat. There are several types of parasites: protozoa are single-celled organisms that are only visible under a microscope, while worms come in all sizes from threadworms that measure less than 1 cm, to tapeworms that grow up to 12 m in length. All these types of parasites grow naturally in many animals such as pigs, cats, and rodents and can be killed by proper cooking and freezing. Parasites enter the food chain through contaminated or unfiltered water, contaminated soil, and contaminated fruits, vegetables, and raw meat.

The summary of common hazards responsible for the safety of food has been shown in Figure 2.2. The presence or absence of these hazards decides whether the food is safe or unsafe.

A microbial contaminant like live bacteria, viruses, fungi, or parasites may cause a foodborne illness. However, illness can be also caused by the biological toxins present in food. Examples are seafood toxins, mushroom toxins, *Clostridium*, and *Botulinum*. Cooking at high temperatures can destroy many microbes but may not

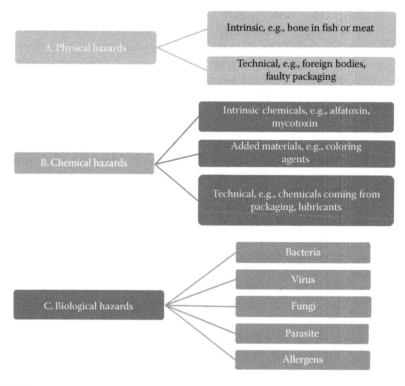

FIGURE 2.2   Common hazards to the safety of food.

be effective in the case of a few toxins as some toxins are thermostable. To prevent biological contaminants, care has to be taken to understand the source of food; foods should only be purchased from reputed companies, and good personal hygiene should be maintained. Proper hand washing, cleaning and sanitization of equipment, and maintenance of cleaning and sanitation facilities are essential to keep food free of microorganisms. Foodborne illnesses are the greatest danger to food safety. Contaminated food can result in illness or disease in an individual that would affect their overall health, work, and personal lives. It can also result in huge financial losses like loss of family income, increased insurance, increased medical expenses, cost of special dietary needs, and loss of productivity, leisure, and travel opportunities. Foodborne illness outbreaks can cost an establishment thousands of dollars and can even be the reason it is forced to close. Loss of customers and sales and loss of prestige and reputation are also major problems. In the case of Nestlé in India, when Maggi (instant noodles) was declared unsafe and banned, it cost millions in the destruction of food products and loss of sales. In the case of foodborne outbreaks, the situation can be dreadful for the company.

Food allergens are one of the naturally occurring substances that can cause the immune system to overreact. Allergens can cause swelling of the lips, tongue, and mouth, difficulty breathing, vomiting, diarrhea, and cramps. Common food allergens that have been identified are milk, soy, egg, fish, wheat proteins, shellfish, peanuts, and chicken. There are other naturally occurring toxins that can cause illness in human beings. Ciguatoxins are responsible for intoxication caused by eating contaminated tropical reef fish. This toxin is found in algae and then eaten by reef fish, which are then eaten by bigger fish such as barracuda, mahi, bonito, jackfish, or snapper, and the toxin is accumulated in the flesh of these fish. The toxin causes nausea, vomiting, diarrhea, dizziness, and shortness of breath. Scombrotoxin causes histamine poisoning, caused by eating food high in a chemical compound called "histamine" that is produced by bacteria. Leaving fish at room temperature usually results in histamine production. It causes dizziness, a burning sensation, facial rash, shortness of breath, and a peppery taste in the mouth. Shellfish toxins are produced by certain algae and are called "dinoflagellates." When eaten by certain shellfish such as mussels, clams, oysters, or scallops, this toxin accumulates in their internal organs and becomes toxic to humans. *Escherichia coli* produces Shiga toxin, which is a poisonous substance. It causes bloody diarrhea followed by kidney failure. Intoxication may be caused by eating undercooked ground beef, unpasteurized apple juice, undercooked fruits and vegetables, and raw milk. Listeriosis is caused by the *Listeria monocytogenes* bacteria, which has the ability to survive in high-salt foods and can grow at refrigerated temperatures. The perfringens foodborne illness is caused by *Clostridium perfringens*. It causes severe abdominal cramps and severe diarrhea. Salmonella bacteria, which are facultative anaerobic bacteria, cause bacterial infection with symptoms like stomach cramps, diarrhea, headache, nausea, fever, and vomiting. Shigellosis bacteria comes from human intestines and polluted water and is spread by flies and food handlers; it causes diarrhea, fever, abdominal cramps, and dehydration. Staphylococcal illness is caused by *Staphylococcus aureus*, which can grow in cooked or safe foods that are recontaminated.

2.4 FOODBORNE OUTBREAK

Foodborne outbreak is an incident in which two or more people experience the same illness after eating the same food. As mentioned earlier, biological hazards are mostly responsible for foodborne outbreaks. Out of these, bacteria are responsible for most of the diseases in humans. Foodborne illness usually arises from improper handling, preparation, or food storage. Good hygiene practices before, during, and after food preparation can reduce the chances of contracting an illness. There is a consensus in the public health community that regular hand washing is one of the most effective defenses against the spread of foodborne illness. The action of monitoring food to ensure that it will not cause foodborne illness is known as food safety. To understand how to control the occurrence of outbreaks, first we need to know the reasons for and contributors to the contamination and spoilage of food. There are six main contributors to the growth of microbes. These are as follows:

1. *Food-to-food contamination*: This occurs when harmful organisms from one food contaminate other foods. For example, raw meat that is thawed on a shelf where it can drip onto other foods will spoil the other foods. In a box of apples or any other fruit, juices dripping out of one spoiled fruit will affect the rest of the fruits as well. Care has to be taken to segregate just the potential spoiler. To prevent food-to-food contamination, store already cooked foods on a higher

shelf of the refrigerator than raw foods. We have to wash fruits and vegetables under cold running water and not prepare raw meat and raw vegetables on the same surface at the same time. Equipment-to-food contamination may be avoided with the use of separate cutting boards for different foods. Prepare raw foods in a separate area from fresh and ready-to-eat foods. Clean and sanitize equipment, work surfaces, and utensils after preparing each food.

2. *Acidity*: Bacteria, which are the main cause of the spoilage of food, grow best in a slightly acidic to slightly neutral environment (pH 4.6–7.5). Only a few bacteria, such as acidophilic bacteria, can develop spores whereby they can grow and multiply in an acidic environment. Normally if the pH is below 4.6 or higher than 7.5, microbes will not grow. Under ideal conditions, bacterial cells can double in number every 25–30 minutes, so foods with the pH range 4.6–7.5 are likely to spoil very quickly compared to more acidic or alkaline foods. The pH range 4.6–7.5 is conducive for the growth of bacteria. Between 7.5 and 9.0 bacteria may survive, but at a pH of less than 4.6 very few microbes can survive. Highly acidic foods such as vinegar and lemons inhibit the growth of microorganisms. Salad dressing made with vinegar, oil, and garlic can stay safe for longer.

3. *Temperature*: The temperature range at which spoilage occurs is high. Temperature abuse happens when the food is exposed to the "temperature danger zone," 41°F–140°F (5°C–60°C), for more than 4 hours. Time temperature abuse occurs when food is not stored, prepared, or held at a required temperature. Food must be cooked or reheated to a temperature high enough to kill harmful microorganisms. If it is cooked and kept at a temperature that only keeps it warm, and it is not cooled to a low enough temperature quickly, this will be conducive to the growth of microbes and thus the food will start to spoil. Food prepared in advance, not consumed immediately, and not set to a safe, required internal temperature will have microbial growth. 41°F–140°F is the danger zone, so food must only be kept at this range for a short amount of time; food should be either cooler than 41°F or hotter than 140°F to prevent harmful microbes from growing. Bacteria can be classified into three categories based on tolerance to temperature: psychrophilic bacteria can grow within the temperature range of 32°F–70°F (0°C–21°C); mesophilic bacteria can grow at 70°F–110°F (21°C–43°C); and thermophilic bacteria grow best above 110°F (43°C).

4. *Time*: As just mentioned, if food is kept for more than 4 hours at a temperature range conducive to the growth of bacteria, it is bound to spoil. Under ideal conditions, with the right temperature and acidity, bacterial cells can double in number every 25–30 minutes. Bacteria multiplies in a logarithmic fashion. If we start with an example of one bacterium, then 30 minutes later it has divided, and we have 2. After 30 more minutes, each of those bacteria have divided, and we have 4 bacteria. Thirty more minutes later we have 8, then 16, and so on. These will continue to grow with each hour, making food unpalatable and unsafe.

5. *Oxygen*: Oxygen is essential for the growth of living beings, including microorganisms. In the absence of oxygen, microbes do not grow. This is

the reason that, to give longer shelf life to foods, vacuum conditions are created in packaging or canning. However, different bacteria differ in their oxygen requirements. Aerobic bacteria need oxygen to grow; most bacteria are aerobic. However, anaerobic bacteria cannot survive when oxygen is present because it is toxic to them. Anaerobic bacteria can survive and grow in vacuum-packaged or canned foods where oxygen is not available. Facultative anaerobic bacteria can grow with or without free oxygen but have a preference for oxygen. Microaerophilic organisms can survive with very little oxygen. Bacteria such as *Clostridium botulinum* and *Clostridium perfringens* live without the presence of oxygen. Since different bacteria grow in different oxygen requirements, it is difficult to control the growth of bacteria by controlling this condition.

6. *Moisture*: Moisture is an important factor in microbial growth. The amount of water available for microbial activity is called "free water" and is measured as water activity. The water activity level is the measure of the amount of water that is available for bacteria to grow, and it is measured from 0 to 1. PHFs that have more free water are foods that have a water activity level of 0.85 or higher. To make water unavailable for microbial activities a few techniques can be adopted: lower the amount of moisture in the food through freezing, dehydrating, or adding sugar or salt.

The previous discussion makes it clear that the foods most likely to become unsafe typically have the following characteristics: a water activity level of 0.85 or more, a pH level of 4.6–7.5, and high protein content. Common foods that fall under this category and are therefore most susceptible to early deterioration are fish, meat (e.g., beef, pork, lamb), milk and milk products, cooked rice, beans, textured soy protein, meat alternatives, poultry, seafood, sprouts, raw seeds, sliced melons, eggs, and baked/boiled potatoes.

2.5   FOOD SANITATION, SAFETY, AND HYGIENE

Since most spoilage is due to the growth of microbes, efforts can be made to keep food in conditions that will eliminate or reduce the chances of microbial growth. Food safety, in simpler terms, is a scientific discipline dealing with the handling, preparation, and storage of food in ways that prevent foodborne illness. This includes a number of routine activities that should be followed to avoid potentially severe health hazards. In fact, most of these precautions, which may appear to be inconsequential, are routine because they are critical to the production of safe food. Before food reaches consumers, it has to travel a long path from the farm stage. The journey of food starts from the sowing of the seed, but care of the safety of the food starts much earlier: the moment the decision is taken to grow it. This is the reason that food safety considerations include the origins of food. Other than that, they include the subsequent stages dealing with the practices related to food labeling, food hygiene, food additives, and pesticide residues, as well as policies on biotechnology and food, and guidelines for the management of governmental import and export inspection and certification systems for foods. The ultimate concern is the safe delivery and preparation of the food for the consumer.

2.5.1 PRINCIPLES OF FOOD HYGIENE

As has been discussed earlier, food can transmit disease as well as serve as a growth medium for bacteria that can cause food poisoning. In developed countries, there are intricate standards for food sourcing and preparation, whereas in underdeveloped countries the main issue is simply the availability of adequate food and safe water. Food poisoning is preventable. The five key principles of food hygiene, according to the World Health Organization (WHO, 2010), are as follows:

1. Prevent contamination of food with pathogens spread by people, pets, and pests. These are the main sources through which microbes are transmitted. Proper sanitation and hygienic practices are adopted to keep food safe.
2. Separate raw and cooked foods to prevent contaminating the cooked foods. While cooking, due to high temperatures, microbes are destroyed. Raw food contains microbes as the food has not been processed. If such raw food comes into contact with cooked food, the microbes will get the opportunity to grow.
3. Cook foods for the appropriate length of time and at the appropriate temperature to kill pathogens. Cooking destroys bacteria and other microbes as they cannot withstand high temperatures for a long time.
4. Store food at the proper temperature. The temperature should be such that it is not conducive to the growth of bacteria. This is the reason that food, especially cooked or processed food, is stored at low temperatures.
5. Use safe water and raw materials. If the microbial load in the raw material is less, it will require less time processing or heating to eliminate the microbes. On the other hand, if the initial microbe population is more, greater efforts will be required to make it free from microbes.

These five principles have been given in Figure 2.3.

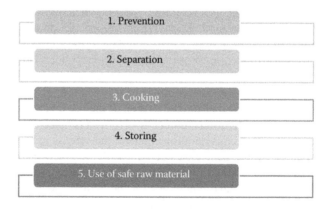

FIGURE 2.3 Key principles of food hygiene.

The availability and consumption of adequate safe and nutritious food is critical to our health, prosperity, and well-being. Keeping our food safe requires an understanding of food safety principles. There are a number of contaminants (or hazards) that could appear in our food, and contamination may occur at any stage during the production, processing, storage, and preparation of our food. Good food safety practices throughout the food production chain, from farm to fork, will minimize the chances of contamination and also minimize, or even eliminate, the impact of contamination that has already occurred. An understanding of these potential food safety hazards reveals the interrelationship between biosecurity, animal health, and food safety. In order to minimize the risk of contamination of poultry, eggs, and meat for consumption, it is important to minimize the risk of contaminants reaching our commercial poultry flocks. Good farm management and hygiene practices and procedures are critical to achieve this objective. These practices will minimize the chance of contamination and will optimize bird health, which will allow the birds' own immune systems to be another safeguard against potential contamination. The process in plants is similar as well. Studies have shown that a large percentage of foodborne illness cases can be attributed to poor sanitation and food hygiene, including poor personal hygiene and contamination of the equipment and/or environment. Examples of food recalls related to sanitation issues include the contamination and subsequent recall of deli meats in Canada in 2008, when cells of Listeria monocytogenes were transferred to the product after surviving in equipment niches where they were protected from sanitation procedures. Food sanitation includes all practices involved in protecting food from risk of contamination by harmful bacteria, poisons, and foreign bodies, and preventing any bacteria from multiplying to an extent that would result in illness in consumers. In its objective to keep food safe, sanitation also involves destroying any harmful bacteria in the food by thorough cooking or processing. The primary objective of food sanitation is the maintenance of absolute cleanliness. This begins with personal hygiene, the safe handling of food during preparation, and clean utensils, equipment, appliances, storage facilities, kitchens, and dining rooms. Control of the microbial quality of food must focus on the preparation of the food itself, food handlers, facilities, and equipment. The quality of food depends on the conditions under which it is purchased, the time and temperature control during storage, preparation, processing, and service. Personal hygiene and the cleanliness of the facilities and equipment contribute to food safety. Sanitation is a process of cleaning to keep the chances of contamination and microbial growth to a minimum, if not totally absent. Cleanliness is the absence of visible soil or dirt but is not necessarily sanitary.

Food handlers can be carriers of disease-causing microbes. Food service personnel can contaminate food in many ways. If any food handler is suffering from an illness, he can pass on the microbes to another person through food. Hepatitis A, Shigella, and *E. coli* infections can spread through food. Cuts and wounds must be kept covered while handling food. Bare hands are covered by gloves while handling ready-to-eat foods. Disposable gloves are used and it is insisted upon that hair be covered, that handlers wear appropriate attire, and that they refrain from wearing jewelry, makeup, and nail polish. Proper hand-washing procedures are followed.

According to the Codex Alimentarius Commission (CAC), food hygiene should cover all of the following elements throughout the supply chain (CAC, 2003b):

1. *Primary production*: During production, environmental hygiene and hygienic production are very important. Sanitary conditions must be maintained during the storage and transportation of material. The maintenance of facility, cleaning, and personnel hygiene is to be ensured.
2. *Establishment*: The design of facilities should be such that it encourages cleanliness. Facilities should not be located near to garbage dumps. Premises and rooms should be well ventilated. Equipment and facilities should be such that they can be cleaned easily.
3. *Control of operation*: There should be a system to identify food hazards and vulnerable points, and hygiene control systems should be activated accordingly. Quality checks have to be done for water and all incoming materials including packaging. The management and supervision of checks and hygiene is very important. Documentation and records of all the processes are required to establish confidence, to identify deficiencies, and also to make corrections to the procedure if any shortcoming is noticed. The organization has to keep recall plans, ready to implement if the need arises.
4. *Maintenance and sanitation of the establishment*: The maintenance and cleaning of the facility and also all the associated equipment has to be undertaken regularly. Sanitation protocols and cleaning programs must be laid down. Pest control systems should be operational. Waste management must not only conform to legal requirements but must also be compliant with environmentally good practice. Monitoring of hygienic practices must be effective.
5. *Personal hygiene*: There should be a health checkup of all the staff in the facility, with special emphasis on the staff handling food. Any illness and injury needs to be reported. Personal cleanliness and personal behavior need to be monitored. Staff must follow the rules and wear the prescribed dress. The same types of rules are applicable to visitors.
6. *Transportation*: The hygiene procedures and sanitation practices followed for the facility are also applicable during the transportation of food material. Product information and consumer awareness (lot identification, product information, food labeling, and consumer education) have also to be taken care of.
7. *Training*: Training creates awareness about the requirements and purpose of hygiene practices among the staff. Responsibilities are to be fixed. Accordingly, training programs are to be designed and detailed instructions are to be issued. Induction training and refresher courses are to be conducted at regular intervals.

Cleaning and food hygiene procedures for the building, plant, and equipment should be validated using visual, analytical, or microbiological methods, and records should be maintained. For instance, swab samples can be taken from various places on equipment, floors, walls, or drains to test for the presence of contamination. Then, after applying a sanitation step, samples can be taken again and compared with the original results to ensure that the step has been effective at reducing harmful microbes to safe levels. For certain high-risk materials (e.g., allergens, ruminant proteins, or ready-to-eat products) validation of procedures is mandated, with individual governments designating acceptable methods for cleaning high-risk materials (CAC, 2009).

2.5.2 Sanitation Facilities and Pest Management

It is important to clean areas and eliminate faulty and not-in-use equipment. It is essential to get rid of dirty surroundings as these will attract bugs and other pests. Facilities for safe food must have walls that are cleaned daily with cleaning solution and floors that are swept and vacuumed daily, and spills should be cleaned immediately. The facility must establish a routine cleaning schedule and should have proper ventilation. Processing areas must have exhaust fans to remove odors and smoke. Always use a hood over cooking areas and dishwashing equipment. Checking of all exhaust fans and hoods must be done regularly to make sure they are working and have been cleaned properly. Hood filters must be cleaned regularly as instructed by the manufacturer. Restrooms should have warm water at 100°F (37°C) for hand washing. Liquid soap, toilet paper, paper towels, or hand dryers should be adequately supplied. Garbage should have a foot-pedal cover and doors should be self-closing. Trash is to be removed daily and a proper system of garbage handling and garbage collection should be maintained. Garbage must be kept away from food preparation areas. Garbage containers must be leakproof, waterproof, pestproof, and durable. Garbage containers should be cleaned and sanitized regularly. Any place that is dark, warm, moist, or hard to clean must be cleaned daily. Holes, boxes, seams of bags, and folds of paper must all be checked daily. Cleanliness and maintenance are key to preventing pest infestation. By its nature, the food service environment is prone to problems with pests. Pests may be brought in when other foods and supplies are delivered. They may also enter the building through gaps in the floors or walls. Every establishment should have an ongoing pest prevention program and regular pest control by a licensed pest control operator. All openings or cracks in walls and floors must be filled. Openings in pipes or equipment fittings must also be filled. Screen all windows, doors, and other outer openings so that insects cannot enter. Food must be kept in containers with tight-fitting lids. Remove and destroy food that is infested. Maintain proper temperatures in storage areas. Dishes, flatware, preparation and serving utensils, measuring devices, and cooking pots and pans must be sanitized and kept clean to remove visible dirt. Ensure the absence of microorganisms by using heat or chemicals (CAC, 2001).

Cleaning agents are chemical compounds that remove food, soil, rust stains, and minerals. There are various types of cleaning agents:

- *Detergents*: All detergents contain surfactants that reduce surface tension between the soil and the surface.
- *Solvent cleaners*: Called "degreasers," solvent cleaners are alkaline detergents that contain a grease-dissolving agent. These work well in areas where grease has been burned on.

- *Acid cleaners*: Used on mineral deposits and other soils, these cleaners are often used to remove scale in warewashing machines.
- *Abrasive cleaners*: These contain a scouring agent that helps scrub off hard-to-remove soil.

Methods in sanitizing: Chemical sanitizing of the material to be sanitized is accomplished by wiping with the sanitizing solution or immersing the material in the sanitizing solution for a specified amount of time. Sometimes equipment is exposed to a high heat for an adequate length of time. This is done manually by immersing equipment into water at 171°F–191°F (77.2°C–88.33°C) for at least 30 seconds.

Most common chemical sanitizers

1. Chlorine is the most commonly used and the least expensive. It is effective in hard water, but it is inactivated in hot water above 120°F (48.89°C). Normally a mixture of chlorine sanitizing solution can be prepared as a 50 parts per million (PPM) solution or a 100 PPM solution as per the requirements.
2. Iodine is effective at low concentrations. It is not quickly inactivated by soil as compared to chlorine. However, it is less effective than chlorine. It also becomes corrosive to some metals at temperatures above 120°F (48.89°C). It is more expensive than chlorine and may also stain surfaces.
3. Quaternary ammonium compounds (quats) are not inactivated as quickly by soil as chlorine and remain active for a short period of time after they have been dried. These are noncorrosive and nonirritating to skin. They work in most temperature and pH ranges. They leave a film on the surface but do not kill certain types of microorganisms.

The most commonly found pests in food processing facilities are cockroaches, flies, beetles, ants, moths, and rodents. These pests create different problems individually and collectively. Due to the operations of any food plant/facility, and also by the nature of these pests, management of pests in food plants can be challenging. Some pests may enter the facility through the delivery of raw food items and packaging materials, or be attracted by the presence of food odors and exterior lighting or the sheer size of the facility, which may provide places to harbor pests: each of these makes the job of handling pests very challenging. The most successful approach to pest management in food plants is the use of integrated pest management (IPM), a process involving comprehensive and coordinated efforts to treat and control pests. This involves three basic steps: inspection, identification, and treatment. The most important objective of sanitation is the prevention and immediate elimination of any and all pest infestations in a food plant. Unsanitary conditions tend to provide the perfect environment for pests by allowing them access to food, water, and nesting sites. Sanitation has to be maintained in any plant, even when not in production. Certain sanitary procedures must be followed and maintained daily to prevent direct product contamination or adulteration of ingredients. Perfect sanitation is essential in these areas to maintain a safe food production process leading to the production of safe food. We may recall the conditions and pest violations inside the Peanut Corporation of America's processing plant that led to a massive *Salmonella* outbreak in 2009. This eventually resulted in the shutting down of the plant, the bankruptcy and liquidation of the company, and judicial action against the management. This example and countless other recalls are the reasons why every food processing plant should develop strict standard operating procedures (SOPs) relating to sanitation and pest management. Because of the crucial role pest management plays in sanitation inside food plants, it is important that each plant develop and establish SOPs, including methods of inspection and treatment, as well as provide appropriate documentation. Any SOP must lay down procedures for all current or anticipated pest management activities, including inspections and audits, and frequency of action. It should also log pest sightings and activity, plant layout, responsible individuals, corrective measures, treatment records, and trend reports.

2.5.3 Principles of Food Facility Design for Proper Sanitation

Food production facilities must be designed in such a way that the chances of contamination are minimized and so that they are, at the same time, easier to clean. The movement of people and material has to be streamlined. These facilities must follow the following broad parameters:

1. *Establishment of distinct hygienic zones in the facility*: Creating hygienic zones through physical separation has become an industry standard and is essential to food safety. It also streamlines the process in the facility. We can keep one line running while another is down for sanitation or changeovers, or use allergens on adjacent lines with no cross-contamination concerns.
2. *Controlled personnel and material flows to reduce hazards*: Once the plant is physically divided into distinct zones, the flow of personnel and materials can be better controlled. Workers, visitors, products, ingredients, and packaging materials must all be carefully managed as they move from one area to another to avoid cross-contamination.
3. *Avoidance of water accumulation inside the facility*: Water is an essential part of products, processes, and sanitation, but it is also one of the greatest threats to food safety. From the walls to the floors and ceilings, the design must prevent accumulation of water and, in fact, facilitate drainage and water removal.
4. *Control of temperature and humidity in the facility*: The facility must have proper fabrication, installation, and servicing of refrigeration and mechanical systems so that the facility is in control of room temperature and humidity to facilitate control of microbial growth.
5. *Airflow and room air-quality control*: Properly controlling the flow of air through a food processing plant is vital to food safety. All air that comes into the main food processing zone must be high-quality, clean air. In addition, air from raw-product zones must travel in the opposite direction and exit directly from the plant.
6. *Building design to facilitate safety and sanitary conditions*: It must be ensured that elements such as security, personnel safety, vehicle management, lighting, grading, and water-management systems facilitate safe, secure, and sanitary site conditions.

7. *Building components and construction to facilitate sanitary conditions*: Food safety can be affected by floor surfaces, wall finishes and coatings, and design items such as walk-on ceilings, so buildings must have seal-tight gaskets around wall-penetrating pipes, and smooth, cleanable finishes on walls and other surfaces that can make the difference in food safety.

8. *Sanitation integrated into facility design*: By properly securing utility systems and specifying materials, finishes, and other sanitary design details in the utility system, design must provide easy access for cleaning, inspection, and maintenance to prevent collection areas for bacteria. Care has to be taken that small things have been ensured, like the design of clean-room entries into processing plants, skid-mounted equipment, round tubing, minimal flat surfaces, use of the right types of chemicals, and the installation of all sinks, conduits, fixtures, and more on stand-offs away from the walls to enable easy access for cleaning. No stone should be left unturned in ensuring a cleanable facility.

2.6 FOOD SAFETY PROGRAMS AND HAZARD ANALYSIS AND CRITICAL CONTROL POINTS

2.6.1 PREREQUISITE PROGRAMS

Prerequisite programs (PRPs) are practices adopted to ensure that the environment is clean, sanitary, and appropriate for manufacturing safe products. Food producers must develop and implement procedures to reduce the potential for contamination with microorganisms and other hazards. The starting point is always the farm, and thus good agriculture practices are the precursor to any activity in food safety. It is extremely important that food product manufacturers develop and implement effective good manufacturing practices (GMPs), good hygienic practices (GHPs), and standard operating procedures (SOPs). PRPs are the foundation of a successful food safety management system. Deficiencies and inadequacies in programs may lead to additional critical control points that would have to be identified, monitored, and maintained. PRPs are the minimum sanitary and processing requirements for the food industry. PRP regulations are designed to control the risk of contaminating foods with filth, chemicals, microorganisms, and other means during their manufacture. PRPs are fairly broad and general and give few details as to what specific procedures must be followed to comply with the regulations. SOPs are the steps the food company takes to assure that the PRPs are met. SOPs are very specific sanitary and processing requirements for the food industry. They include stepwise procedures, employee training, monitoring methods, and records used by the food industry. PRPs are not designed to control specific hazards but are intended to provide guidelines to help food processors produce safe and wholesome products. PRPs and SOPs are the foundation of the food safety programs of any food production plant. PRPs can be used to help guide the development of SOPs.

PRPs are basically guidelines for personnel, buildings and facilities, equipment and utensils, and production and process control to provide a safe practice for manufacturing, packing, or holding food for human consumption. To implement PRPs, the following steps are required to be taken:

1. *Standard fixing*: Select the correct prerequisite standard document. This will be specific to the product, process, and facility.
2. *Gap analysis*: A gap analysis is a prerequisite to evaluate the current system and practices against the requirements of the standard. Gap analysis will give details of actual requirements.
3. *Plan of action*: Document current PRPs, if any, and address any gaps found during the gap analysis. This will be a plan for activities that will have to be undertaken.
4. *Role assignment*: Assign responsibilities and define roles for implementing the PRPs.
5. *Verification and validation*: This is essential to see if the action taken is adequate and sufficient to address the food safety issue.

2.6.2 Good Manufacturing Practices

Good manufacturing practice (GMP) is a system for ensuring that products are produced consistently and controlled according to quality standards. GMP can be classified into two categories. There are general GMPs that lay down the Good Manufacturing Practices, which are common to all products. For specific products there will be specific requirements, so the GMPs will be specific. According to EU Regulation (EC) 2023/2006, GMP means those aspects of quality assurance that ensure that materials and articles are consistently produced and controlled to ensure conformity with the rules applicable to them and with the quality standards appropriate to their intended use by not endangering human health or causing an unacceptable change in the composition of the food or causing a deterioration in the organoleptic characteristics thereof.

GMP should be able to achieve certain levels of uniformity and consistency in operations. It is designed to minimize the risks involved in any food production that cannot be eliminated through testing the final product. The main risks are: unexpected contamination of products, causing damage to health or even death; incorrect labels on containers, which could mean that consumers receive the wrong food; and insufficient or too much additive or contaminant, resulting in ineffective preservation or unsafe food. GMP should ensure that quality is constantly monitored by observing the following:

- *Size of facility*: Food production facilities and processing areas are of the appropriate size to prevent overcrowding and to allow proper placement and orderly handling of equipment, raw materials, and other product materials such as packaging and labeling.
- *Raw material quality*: All the raw material used, and ingredients including food additives, are of the appropriate level of quality and safety.
- *Segregation of material*: Storage is proper to prevent contamination and mix-up with other processing material.
- *Facilitative layout*: The layout of facilities permits the orderly flow of production materials and personnel in processing.

- *Proper light and ventilation*: Facilities are suitably lit and ventilated for proper light and air circulation.
- *Proper maintenance*: Equipment and all other facilities are maintained for proper functioning.
- *Proper control of parameters*: Temperatures, times, pressures, machine operations, and other processing parameters are controlled at the specification level required to assure proper processing.
- *Labeling*: Appropriate labels are used for each product and at each level of processing.
- *Testing and control procedures*: The examination/testing or sampling of intermediate foods from the processing lines and finished foods from final storage. The products are examined or tested analytically for compliance with product specifications and quality and safety requirements.

GMPs are enforced in the United States by the U.S. Food and Drug Administration (FDA), under Title 21 CFR. The regulations use the phrase "current good manufacturing practices" (cGMP) to describe these guidelines. Since June 2010, a different set of cGMP requirements has been applied to all manufacturers of dietary supplements. The WHO's version of GMP is used by pharmaceutical regulators and pharmaceutical industries in over 100 countries worldwide, primarily in the developing world. The European Union's GMP (EU-GMP) enforces similar requirements to the WHO's GMP, as does the FDA's version in the United States. Similar GMPs are used in other countries, with Australia, Canada, Japan, Saudi Arabia, Singapore, the Philippines, Vietnam, and others having highly developed/sophisticated GMP requirements. All these regulations on GMP have a monitoring mechanism through the maintenance of records. The establishment of recordkeeping systems for recording the results of quality control activities at all levels of processing is an integral part of GMP. Information that might be recorded includes

- Results of quality assurance inspections of production facilities prior to and during production
- Processing parameters during food processing (e.g., cooking times, temperature recordings, pressures)
- Results of specific methods or procedures for online product examination (e.g., net weights, can seal tear-down)
- Results of examination of the integrity of the package closure systems
- Specific laboratory analysis methods to be used for quality and safety determinations, sample size, and established criteria for acceptance or rejection of the lot

2.6.3 Good Hygienic Practices

Since cross-contamination of food by food handlers has been a major challenge, special care has to be taken to address this issue. The hygienic conditions and habits of workers determine the amount of cross-contamination from workers to food products. Clean, disease-free, and habitually hygienic persons are necessary to produce clean and safe food products. The General Principles of Food Hygiene provide the basis for food hygiene and lay a firm foundation for the development of an effective system of Hazard Analysis and Critical Control Points (HACCP) or equivalent. The application of these general principles and of GMPs allows the producer to operate within environmental conditions favorable to the production of safe food. The General Principles of Food Hygiene follow the food chain from primary production through to final consumption, highlighting the key controls at each stage. In brief, they give guidance on the design and facilities of premises, in-process control, required support programs of sanitation and personal hygiene, and consideration of hygiene controls once the product has left the production premises. The controls described in the General Principles of Food Hygiene are internationally recognized as essential to ensuring the safety and suitability of food for consumption. The general principles recommended to governments, industry, and consumers are similar. The industry includes individual primary producers, manufacturers, processors, food service operators, and retailers. All collectively have a responsibility to ensure safe food for the consumer and reduce the incidence of foodborne illness and food spoilage.

2.6.3.1   Sanitary Standard Operating Procedures

Implementation of GHPs entails the use of appropriate cleaning and sanitary measures to prevent microbial contamination, and assurance of optimum sanitary conditions for processing food products. Sanitary standard operating procedures (SSOPs) involve:

- Use of appropriate cleaning and sanitizing techniques, including the use of approved and effective agents at the proper levels (i.e., strength, concentration, temperature, time) and frequency to prevent microbial buildup on processing equipment and utensils or other food contact surfaces
- Procedures to identify and document problems and their corrective actions
- A list of staff responsible for inspecting and verifying that the sanitation is acceptable
- Observation of sanitary practices, use of protective clothing, and strict observation of rules on personal hygiene by personnel involved in handling and processing food
- Use of hand-washing and hand-sanitizing dip stations when and where appropriate
- Having time and temperature controls in place to prevent microbial growth in the susceptible intermediate and finished processed foods
- Use of other sanitary measures that are specifically needed because of the nature of the food being processed, the processing technology, or the facilities in which the processing takes place
- A record of the daily facility/equipment sanitation checks
- Documented verification to reflect changes in equipment and facilities, processes, new technology, or designated establishment employees

2.6.3.2   Preoperational SSOPs

Preoperational SSOPs are established procedures that describe the daily, routine sanitary procedures that occur before processing begins. The procedures must include the cleaning of product contact surfaces of facilities, equipment, and utensils to prevent direct product contamination or adulteration. These might include:

- Descriptions of equipment disassembly and reassembly after cleaning, use of acceptable chemicals according to label direction, and cleaning techniques
- Application instructions, including concentrations for sanitizers applied to product contact surfaces after cleaning

2.6.3.3   Operational SSOPs

Operational SSOPs are established procedures that describe the daily, routine sanitary procedures that will be conducted during operations to prevent direct product contamination or adulteration. Established procedures for operational sanitation must result in a sanitary environment for preparing, storing, or handling any meat or poultry food product. Established procedures during operations might include, where applicable:

- Equipment and utensil cleaning/sanitizing/disinfecting during production, as appropriate, on breaks, between shifts, and in mid-shift cleanup
- Procedures for employee hygiene, such as cleanliness of outer garments and gloves, hair restraints, hand washing, health, and so on
- Product handling in raw and cooked product areas

2.6.4   HAZARD ANALYSIS AND CRITICAL CONTROL POINTS

In the early 1960s, a collaborative effort between the Pillsbury Company, NASA, and the U.S. Army Laboratories began with the objective to provide safe food for space expeditions. In order to ensure that the food sent to space was safe, strict microbial requirements and pathogen limits were established. Using the traditional end product testing method, it was soon discovered that almost all of the food manufactured was being used on testing and very little was left for actual use. A new approach—prevention rather than inspection—was needed. Pillsbury's training program for the FDA in 1969, titled "Food Safety through the Hazard Analysis and Critical Control Point System," was the first time that the HACCP approach was used (Sperber and Stier, 2010). Since then, HACCP has been recognized internationally as a logical tool for adapting traditional inspection methods to a modern, science-based food safety system based on risk assessment, and it has spread from the farm to the fork in all realms of the food industry from small to large enterprises (FAO/WHO, 2007). In the manufacturing and processing of foods it is also necessary to ensure the safety of ingredients used as technical aids, additives, flavorings, or colorings. Such safety assessments require the analysis of test data, chemical specifications for substances involved, and information on human dietary consumption levels and patterns. It is also necessary to evaluate the impact of uncertainties in cases where the information is insufficient to make a clear safety assessment decision. This is the work of highly trained specialists in toxicology, nutrition, chemistry, food composition, and risk assessment techniques. The necessary expertise is often found only in countries with highly trained personnel and advanced technological capabilities (CAC, 2003a).

HACCP is a systematic, preventive approach to food safety and allergenic, chemical, and biological hazards in production processes that can cause the finished product to be unsafe; the HACCP system designs measurements to reduce these risks to a safe level. In this manner, HACCP is referred to for the prevention of hazards rather than for finished product inspection. The HACCP system can be used at all stages of a food chain, from food production and preparation processes including packaging, distribution, and so on. However, in order for the HACCP system to be effective, there must first be an effective PRP system in place. Combining strong PRPs with HACCP will increase the total process control system and help food product manufacturers continue to produce the safest products possible. The development and successful implementation of these programs requires full management support and commitment (International HACCP Alliance).

HACCP includes the identification of all the known potential hazards that can be associated with the food being processed. Once this hazard assessment is done, CCPs in the processing are identified where controls can be exercised to prevent, reduce, or eliminate these hazards. Constant vigilance is maintained over the CCPs to prevent any process deviations that would result in loss of control at the CCP. Appropriate corrective actions are required whenever a CCP is found to be out of control, and the suspect food product is generally prevented from being distributed until its safety and acceptability have been determined. This system is highly effective when employed properly, but it requires considerable understanding and technical information related to the food product, the processing methods, and the production facility.

2.6.4.1 Principles of HACCP

2.6.4.1.1 Principle 1: Conduct a Hazard Analysis and Risk Assessment

All reasonable hazards must be determined at this step. A food safety hazard is any biological, chemical, or physical property that may cause a food to be unsafe for human consumption. To determine a complete list of hazards, a brainstorming session should be organized. It is important that all hazards are precisely stated. The HACCP team should next conduct a hazard analysis, which includes a risk assessment to identify which hazards are of such a nature that their elimination or reduction to acceptable levels is essential to the production of safe food. The risk assessment is a combination and result of the likelihood of occurrence and the severity (i.e., the degree of consequences for health) to establish the degree of concern. For example, a piece of glass in a food product has a low risk but a high severity. Next to the identification of each hazard, all control measures are taken within the process to prevent the possible occurrence of each hazard, to eliminate the hazard, or to reduce the hazard to an acceptable level.

2.6.4.1.2 Principle 2: Identify Critical Control Points

A critical control point (CCP) is a point, step, or procedure in a food manufacturing process at which control can be applied and, as a result, a food safety hazard can be prevented, eliminated, or reduced to an acceptable level. To identify the CCPs, the so-called CCP decision tree is used. All hazards that have been determined must be evaluated by this decision tree. Note that the decision tree is just a tool to be used. Common sense and discussion among the HACCP team members will eventually decide whether or not the process step is a CCP. Make sure that the arguments in the discussion are registered.

2.6.4.1.3   Principle 3: Establish Critical Limits for Each Critical Control Point

The team must determine measurements or values that indicate the operational limits for each critical limit (CL). These values represent the set values, targets or norms, target levels, tolerances, control limits, warning levels or action limits, and critical limits that are appropriate for the effective elimination of the hazard, and these should be established and specified. This may require some investigation. Quantifying the targets, target levels, and CLs will not always be easy (e.g., how to quantify the effectiveness of cleaning operations).

2.6.4.1.4   Principle 4: Establish Critical Control
Point Monitoring Requirements

Monitoring activities are necessary to ensure that the process is under control at each CCP. The principle describes the techniques and methods of measurement used to evaluate CLs to ensure that CLs are not infringed upon and targets are maintained. Continuous monitoring is preferred. The process should be monitored at the most likely point of infringement.

2.6.4.1.5   Principle 5: Establish Corrective Actions

Corrective actions are actions to be taken when monitoring indicates a deviation from an established critical limit. If the critical limit of the product is not met, it requires the food processing plant's HACCP plan to identify the corrective actions. Corrective actions are intended to ensure that no product injurious to health or otherwise adulterated as a result of the deviation enters commerce. Corrective actions must be identified for each CCP, stating who, when, and how they should act, and last but not least what will be done with the affected product. Systems must be developed to deal with affected product should CLs be infringed. To register the nonconformance incident, apply the procedure "Production Non-Conformance Report." Other compulsory procedures are the "Consumer Complaints Procedure," "Traceability Procedure," and "Product Recall Procedure."

2.6.4.1.6   Principle 6: Establish Procedures for Ensuring the
HACCP System Is Working as Intended

It is important to verify that the HACCP plan is operating effectively and according to documented procedures (compliance). Verification ensures that the food processing plants do what they were designed to do to ensure the production of a safe product and, if not, what necessary actions should be taken to correct the situation. These verification activities must be carried out periodically in a planned manner to ensure that HACCP plans are maintained and effective. Verification procedures may include such activities as reviewing HACCP plans, CCP records, CLs, and microbial sampling and analysis, but also reviewing the results of various audits, consumer complaints, and nonconformance reports. The HACCP plan includes verification tasks to be performed by plant personnel. Verification tasks would also be performed by inspectors. Undertaking microbial testing is one of several verification activities. Verification also includes "validation"—the process of finding evidence for the accuracy of the HACCP system. The review of the HACCP plan (reassessment) is used to determine whether the plan is still appropriate to the process of verification. Changes in the process, routing, layout, and modification to processing equipment will have to be adapted in the HACCP system. The frequency of reviewing is laid down in the "Procedure of Verification."

2.6.4.1.7   Principle 7: Establish Recordkeeping Procedures

The Codex Alimentarius recommends the assembly of all HACCP documentation and procedures in a manual. Proper documentation control and an appropriate procedure are required to ensure that the data remains up-to-date. All food plants maintain certain documents, including the hazard analysis and written HACCP plan, and records documenting the monitoring of CCPs, CLs, verification activities, and the handling of processing deviations. This requires documented instructions (SOPs) and accurate recordkeeping for all CCPs.

2.6.4.2   HACCP Implementation

is a step-by-step system. In all, there are 12 steps required to develop and implement a HACCP plan. These 12 steps can be further divided and clubbed into two groups: preparatory steps and implementation steps (CAC, 2003a).

2.6.4.2.1   Preparatory Steps

Most HACCP experts believe that a company will do a better job of HACCP plan development if it takes some preliminary steps before it attempts to apply the seven principles and write a plan. A food processing company should take the following five steps to get started.

Step 1: Form a HACCP team and define the scope of the HACCP plan: The team must provide expertise in food safety risks and risk management as well as the day-to-day operations of the business. The team should be relevant, knowledgeable, and multidisciplinary. HACCP is an overall process control system, and it takes a variety of different kinds of knowledge and experience to develop a good system. Using a blend of production and management personnel will ensure that "shop floor" ideas, issues, and observations are developed into a functional HACCP plan. This team could include

- *A team leader*: A communicator who is well informed of HACCP principles and practices
- *Production personnel*: Personnel who are acquainted with the issues, needs, and potential hazards
  - *Quality assurance (QA)/Quality control (QC) personnel*: Personnel who are invaluable in creating compliant documentation and procedures. They will be essential in establishing CCPs and helping to identify areas of risk. Laboratory analysis will be required to establish the

conditions that control (or fail to control) potentially pathogenic organisms in the food production process, particularly if there is no established technical literature.

- *A process engineer*: A person who will determine the mechanical/processing risks and the means of controlling potential risks. The process engineer will be aware of the functional limitations of specific process limitations and will be invaluable in creating the essential process flow chart.

Step 2: Describe the food and its method of production and distribution: Description of the product is essential to fully understand potential risks and control issues. Description of the product must be comprehensive and should include descriptions of

- Identity (e.g., product name and raw materials)
- Factors impacting safety (e.g., microbial loading, water activity, pH, temperature of processing and/or storage, etc.)
- Chemical composition
- Physical structure
- Processing
- Required shelf life
- Packaging and packaging materials
- Storage and handling
- Distribution and environmental conditions
- Labeling and instructions

Step 3: Identify the intended use and consumers of the products: Fully understanding the intended (consumer) usage of the product is a key requirement of the HACCP team. Consumer handling and usage may greatly influence both the quality and safety of the food product. Many food products are susceptible to microbial and/or pathogenic contamination. How the product is to be used may have a direct impact on the required degree of food safety and identification of CCPs. Target groups who can be particularly vulnerable (e.g., the elderly or infants) must be identified for certain products.

Step 4: Develop a process flow diagram: A "flow diagram," also called a "process diagram" or "flow chart," is intended to provide a clear, precise visualization of all the steps involved in the production process. There are no rules for the presentation of flow charts, except that each process step and the sequence of the steps should be clearly outlined. This must include initial handling and receipt of the raw materials, the product flow, timings, temperatures, and all other conditions/steps involved in the process. Areas of specific interest include the interaction between packaging and raw materials, environmental conditions at each stage of the process, as well as the impact of cleaning and preventive maintenance cycles on the overall process. To simplify communication with third parties (e.g., audits), it is recommended to use international standards in a flow diagram. It has also been found to be very useful to make a small description of each process step and at least explain the objective of the step. This information has to be agreed upon by the HACCP team. At a later stage, while applying the decision tree, the objective of the process step will already be laid down and thus unnecessary discussions will be avoided. A floor plan should be drawn. The initial steps in HACCP analysis involve identifying specific hazards (e.g., staff and their contamination points in the factory).

Step 5: Verify the process flow diagram on-site: Steps must be taken to confirm the processing operation against the flow diagram during all stages and hours of operation and to amend the flow diagram where appropriate. The confirmation of the flow diagram should be performed by a person or persons with sufficient knowledge of the processing operation.

Step 6: List all the potential food safety hazards: List all the potential food safety hazards associated with each stage of the production process, conduct a hazard analysis to determine the likelihood (the probability of the occurrence of a hazard, e.g., the probability that it will occur every time or just one time in a million) and potential severity (the magnitude of a hazard, e.g., how many people will get sick or what the liability costs will be) of all hazards and consider any control measures. Each hazard is rated using a system that suits your products. This may be a simple system such as a "low–high" rating. As a general rule, hazards that involve a low likelihood and low severity and are easily controlled are usually not included in a HACCP plan. An overall risk rating is therefore given to assist in determining what hazards to include. In conducting the hazard analysis, wherever possible the following should be included:

- The likely occurrence of hazards and the severity of their adverse health effects.
- The qualitative and/or quantitative evaluation of the presence of hazards.
- The survival or multiplication of microorganisms of concern.
- The production or persistence in foods of toxins, chemicals or physical agents and the conditions leading to them. Consideration should be given to what control measures, if any, exist and how they can be applied to each hazard. More than one control measure may be required to control a specific hazard and more than one hazard may be controlled by a specified control measure. The ratings can be as follows:

Severity	Likelihood	Overall Risk Rating
Low	Low	Low
High	Low	High
Low	High	High
High	High	High

All hazards that receive a high overall risk rating must be controlled in the HACCP plan.

Step 7: Determine the critical control points: These are points at which the hazard must be controlled to prevent a food safety risk in the final product. There may be more than one CCP to which control is applied to address the same hazard. The determination of a CCP in the HACCP system can be facilitated by the application of a decision tree (Figure 2.2), which indicates a logical reasoning approach. The application of a decision tree should be flexible, depending on whether the operation is for production, slaughter, processing, storage, distribution, or other. It should be used for guidance when determining CCPs. However, the example of a decision tree may not be applicable to all situations; other approaches may be used. Training in the application of the decision tree is recommended. If a hazard has been identified at a step where control is necessary for safety, and no control measure exists at that step or any other, then the product or process should be modified at that step, or at any earlier or later stage, to include a control measure. There will also be steps in the process where control is desirable but not critical; these are known as control points.

Step 8: Establish critical limits for each critical control point: Critical limits must be specified and validated for each CCP. In some cases more than one critical limit will be elaborated at a particular step. Criteria often used include measurements of temperature, time, moisture level, pH, water activity, available chlorine, and sensory parameters such as visual appearance and texture. Where HACCP guidance developed by experts has been used to establish the critical limits, care should be taken to ensure that these limits fully apply to the specific operation, product, or groups of products under consideration. These critical limits should be measurable.

Step 9: Establish a monitoring system for each critical control point: Monitoring is the scheduled measurement or observation of a CCP, relative to its critical limits. The monitoring procedures must be able to detect loss of control at the CCP. Further, monitoring should ideally provide this information in time to make adjustments to ensure control of the process to prevent violating the critical limits. Where possible, process adjustments should be made when monitoring results indicates a trend toward loss of control at a CCP. The adjustments should be taken before a deviation occurs. Data derived from monitoring must be evaluated by a designated person with the knowledge and authority to carry out corrective actions when indicated. If monitoring is not continuous then the amount or frequency of monitoring must be sufficient to guarantee the CCP is in control. Most monitoring procedures for CCPs will need to be done rapidly because they relate to online processes and there will not be time for lengthy analytical testing. Physical and chemical measurements are often preferred to microbiological testing because they may be taken rapidly and can often indicate the microbiological control of the product. All records and documents associated with monitoring CCPs must be signed by the person(s) doing the monitoring and by a responsible reviewing official(s) of the company.

Step 10: Establish corrective action plans: Specific corrective actions must be developed for each CCP in the HACCP system in order to deal with deviations when they occur. The actions must ensure that the CCP has been brought under control. Actions taken must also include the proper disposition of the affected product. Deviation and product disposition procedures must be documented in the HACCP recordkeeping.

2.6.4.2.2   Implementation Steps

Step 11: Establish verification procedures: Verification and auditing methods, procedures, and tests, including random sampling and analysis, can be used to determine if the HACCP system is working correctly. The frequency of verification should be sufficient to confirm that the HACCP system is working effectively. Verification should be carried out by someone other than the person who is responsible for performing the monitoring and corrective actions. Where certain verification activities cannot be performed in-house, verification should be performed on behalf of the business by external experts or qualified third parties. Examples of verification activities include

- Review of the HACCP system and plan and its records
- Review of deviations and product dispositions
- Confirmation that CCPs are being kept under control

Where possible, validation activities should include actions to confirm the efficacy of all elements of the HACCP system.

Step 12: Establish documentation and recordkeeping: Efficient and accurate recordkeeping is essential to the application of a HACCP system. HACCP procedures should be documented. Documentation and recordkeeping should be appropriate to the nature and size of the operation and sufficient to assist the business to verify that the HACCP controls are in place and being maintained. Expertly developed HACCP guidance materials (e.g., sector-specific HACCP guides) may be utilized as part of the documentation, provided that those materials reflect the specific food operations of the business. Documentation examples include

- Hazard analysis
- CCP determination
- Critical limit determination

Record examples include

- CCP monitoring activities
- Deviations and associated corrective actions
- Verification procedures performed
- Modifications to the HACCP plan

An example of a HACCP worksheet for the development of a HACCP plan is shown in Figure 2.3. A simple recordkeeping system can be effective and easily communicated to employees. It may be integrated into existing operations and may use existing paperwork such as delivery invoices and checklists to record, for example, product temperatures.

2.6.4.2.3 Additional Steps

Step 13: Train staff for HACCP implementation: Training of personnel in industry, government, and academia in HACCP principles and applications and increasing the awareness of consumers are essential elements for the effective implementation of HACCP. As an aid in developing specific training to support a HACCP plan, working instructions and procedures should be developed that define the tasks of the operating personnel to be stationed at each CCP. Cooperation between primary producers, industry, trade groups, consumer organizations, and responsible authorities is of vital importance. Opportunities should be provided for the joint training of industry and control authorities to encourage and maintain a continuous dialogue and create a climate of understanding in the practical application of HACCP.

Step 14: Maintain and improve the HACCP plan: HACCP plans must be updated with any changes to the operations of a business or to rectify any problems identified through the monitoring or verification procedures.

There are a number of ways of implementing these seven principles into a functional HACCP plan. One implementation technique has been described in Section 2.6.4.2 as the 12 steps, as illustrated in the diagram. Some HACCP experts like Kit Fai Pun and Patricia Bhairo-Beekhoo would like to add two more steps, notably "define the terms of reference" as a first preliminary step, "review the HACCP study" as a final step, and combine Steps 4 and 5, making it a 14-step plan.

In the United States, regulations demand a preliminary step to decide whether meat and poultry production can be grouped according to the following mentioned nine process categories (together with some examples):

- Slaughter, all species: Beef, swine, and poultry
- Raw product, ground: Ground beef, ground pork, ground turkey
- Raw product, not ground: Boneless cuts, steaks
- Thermally processed, commercially sterile: Canned beef stew, pasta with meat
- Not heat treated, shelf stable: Dried salami
- Heat treated, shelf stable: Beef jerky
- Fully cooked, not shelf stable: Hot dogs, wieners, roast beef, ham
- Heat treated but not fully cooked, not shelf stable: Partially cooked patties, bacon
- Product with secondary inhibitors, not shelf stable: Corned beef, cured beef tongue

After completing the preliminary steps, the seven principles of HACCP are applied and HACCP plan is developed.

Some HACCP experts would include "review existing food safety PRPs and meet the requirements for GMPs and SSOPs" as a preliminary step. In any case, it is vital before implementing any kind of food safety program to review an existing program or implement a proper PRP.

2.6.4.2.4 Problems Occurring during Implementation
Potential problems during implementation are the following:

- HACCP is introduced before GMP. If requirements related to GMP are not attained at an appropriate level, then it will be impossible to control all hazards that may occur, due to the inadequate situation.
- The required time, resources and/or knowledge are underestimated.
- HACCP concepts are not fully understood.
- Changes in processing conditions are ignored.
- Unimportant hazards are identified and too many CCPs are introduced.
- What goes wrong may be identified as the hazard, rather than the hazards it causes.
- Flowcharts are too complex.

- HACCP does not have sufficient commitment from the management.
- HACCP is set up without involvement and training of personnel.
- The wrong person is appointed as chairperson of the HACCP team.

There are a few precautions that can be taken to make the exercise successful:

- Total management commitment
- Application of more than one flowchart, if necessary, by making use of connectors, rather than making complex flowcharts
- Use of multidisciplinary groups rather than one permanent team
- Establishment of a "Time & Event Schedule" to be used as a tool throughout implementation
- Development of procedures, working instructions, and checklists in an effective and user-friendly way
- On-time training of staff and employees, especially regarding hygiene measures and GMP
- Accountability and responsibility of the production employees for what they are doing, and giving them the necessary authorization to handle situations accordingly
- Encouragement and/or implementation of in-line process control to be carried out by the production employees
- Appointment of an appropriate person as chairman of the HACCP team
- Specific attention paid to downtime, rework, and waste material

2.6.5 TOTAL QUALITY MANAGEMENT

An essential foundation of any activity involved in food manufacture, handling, and catering is a thorough understanding of the appropriate requirements of GHP and GMP associated with the particular product or commodity area. The adherences to these good practices are the absolute minimum in any food business.

HACCP is now widely adopted as an essential approach to the systematic identification of hazards associated with the manufacture, distribution, and use of food products. It provides a mechanism for defining preventive measures for hazard control. While GMP and GHP address the generic requirements for manufacturing safe food, the benefit of HACCP is that it addresses specific determinants unique to a particular product and process.

Most food businesses have a quality system that addresses all aspects of quality control and assurance. There are many forms of such systems; perhaps the most widely utilized is that based on the ISO 9000 series of standards. Where such a system exists, HACCP is an integral part of the overall quality system.

From Figure 2.3, it can be appreciated that food safety management is an integral part of overall quality management activity and a key component of the longer-term managerial strategy to further enhance the safety and quality of products.

The tools used in an integrated approach for the management of product quality have been described in previous chapters. They comprise the use of those elements of GMP that are specifically concerned with the generic design and operation of hygienic premises and equipment, and the hygiene of personnel and their training; those GMP requirements associated with hygiene (i.e., GHP requirements) form the basis for the operation of a hygienic food operation. The product quality requirements for the manufacture of a specific foodstuff, and the design and operation of the associated plant such that production of safe food is assured, can only be established by applying HACCP; this is the second of the tools described. The importance of applying HACCP within a framework of GMP is emphasized, recognizing the importance of both GMP/GHP and HACCP in assuring product safety.

The third tool is the use of a quality management system, such as the ISO 22000 FSMS 2005, as a means of effectively managing total product quality, recognizing that the control procedures established in a HACCP plan fit well into such a management system and can be readily incorporated into it. The fourth tool, total quality management (TQM) embraces quality, productivity, and safety at the operational and strategic levels as a means for generating better commitment from all members of an organization to achieving these aims. TQM will provide added confidence that products will conform to quality needs. It is of the greatest importance to understand that quality assurance is an indispensable part of the process of product innovation and that it helps in assuring that new products are safe and at the same time meet consumer demands.

2.7   FOOD TRACEABILITY

Despite the fact that there are food safety systems in place, that tough regulations have been enacted, and that improved technology is being used in food processing, we are faced with an increasing number of well-publicized incidences of contamination in food supply, such as *E. coli* in beef, mad cow disease, melamine in milk, and numerous occurrences of pathogenic bacteria like *Salmonella* and *Listeria*. Incidents like these have raised awareness of the need to ensure food quality and safety. Consumers are interested to know the origins of food, the processes it has undergone, and to understand the declarations on the label. It is true that food production and distribution systems are becoming increasingly interdependent, integrated, and globalized. It is not easy to know the reason for the presence of any ingredient or contaminant in food. The supply chain has become long and complicated, but whatever the difficulties, the answers to all these questions must be found in order to establish confidence in the safety of the food. A system or mechanism to trace food consistently and efficiently from the point of origin to the point of consumption can be the answer to these and other questions about food safety in the supply chain. For monitoring compliance with quality, safety, product attributes, and the process confirmation of food, traceability is increasingly becoming an integral element of public and private systems. Traceability is a responsibility shared by all actors in the food chain, ranging from farmer to processor. It is important to integrate small-scale agricultural producers and food processors into the food chain, and to build their capacity to implement traceability systems.

Food traceability refers to recordkeeping systems that provide the ability to identify the path and the history of an animal, food product, food ingredient, or feed through the food supply chain. The CAC (2006) defines traceability as "the ability to follow the movement of a food through specified stage(s) of production, processing, and distribution." The traceability/product-tracing tool should be able to identify at any specified stage of the food chain (from production to distribution) where the food came from (one step back) and where the food went (one step forward), as appropriate to the objectives of the food inspection and certification system. According to the EU Food Law, "Traceability is the ability to trace and follow a food, feed, food-producing animal or substance intended to be, or expected to be incorporated into a food or feed, through all stages of production, processing and distribution." Here food traceability applies to all companies involved in the food supply chain (European Commission, 2007).

2.7.1 Uses of Traceability

Traceability serves many purposes:

1. *Fresh produce traceability for quality control*: Fresh produce must move quickly through the supply chain to avoid spoilage. After harvest, fresh produce is handled and packed by a shipper or by a grower-shipper and exported or sold directly or through wholesalers and brokers to consumers, retailers, and food service establishments. Traceability systems track fresh produce along the supply chain to identify sources of contamination, monitor cold chain logistics, and enhance quality assurance.
2. *Bulk produce traceability for product authenticity*: Bulk produce is more challenging to trace than fresh produce. Products such as grain, coffee, olive oil, rice, and milk from multiple farms are combined in silos and storage tanks, making it difficult to trace them back to their sources (IFT, 2009). Still, overcoming these difficulties, traceability is used for bulk produce as well.
3. *Traceability for safety and sustainability*: Seafood traceability enhances the value of suppliers' brands and consumers' confidence in those brands. For traceability, monitoring, and control, data about the farm of origin, processing plant, current location, and temperature are collected and made available to participants in the supply chain, including wholesalers, shippers, and retailers.
4. *Traceability for disease control and product safety*: Unlike other food industries, the livestock industry has a long history of implementing animal identification and traceability systems to control disease and ensure the safety of meat and dairy products. Lessons from livestock traceability systems may apply to other areas of food safety.

2.7.2 Principles of Food Traceability

The basic aim of any traceability system is that it should be able to track the supplier, process, and customer. This is only possible if there is efficient documentation and recordkeeping. The system should be able to identify and trace the material supplied, when and how it was processed/handled, and where the finished product was sent. The seven principles of an effective traceability system, as given in Figure 2.4, can be

1. *Verifiable*: The system should be verifiable and, whenever back inquiry is made, the origin and movement should be traceable. Traceability is a way of responding to potential risks that can arise in food and feed supply chains to help assure the quality and safety of the food supply. In case a risk is identified, a traceability system enables the national authorities or food businesses to trace the food back to its source in order to swiftly isolate the problem and prevent contaminated products from reaching consumers. Traceability also pinpoints underlying causes of contamination.
2. *Applied consistently and equitably*: The system should work in all situations at all times. Traceability enables companies to have precise, detailed historical data about the timing, handling, condition, and flow of goods.
3. *Result oriented*: Whatever is being claimed must be confirmed. A freshness guarantee is possible when temperature and humidity are recorded from harvest to final sale. Some consumers will pay a premium for products that demonstrate social responsibility traits or health benefits (e.g., organic, natural, etc.), thus requiring traceability.
4. *Cost-effective*: The cost of traceability should not be prohibitive. Through effective communication across the food supply chain, traceability helps identify different risks associated with a food supply chain.
5. *Practical to apply*: The solution suggested should be easily done and serve the purpose. It should be easily understood and followed.

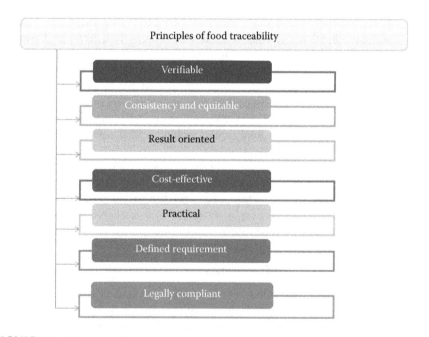

FIGURE 2.4 Principles of food traceability.

6. *Defined in terms of requirements*: Traceability allows for targeted withdrawals/recalls and the provision of accurate information to the public, thereby minimizing disruption to trade.

7. *Compliant with regulations*: Traceability systems should conform to the provisions of food law.

2.7.3 ESSENTIAL COMPONENTS IN FOOD TRACEABILITY SYSTEMS

Traceability systems have become important management information systems within food production and marketing companies. These can efficiently control the supply chain by minimizing the risks of the production process, and help to enhance consumer/customer reliability on products. To achieve this, the following three components have major roles to play:

1. *Identification of premises*: A premises is a parcel of land defined by a description, on which processing and production activities take place. A premises includes the land and associated structures. Accurate identification of premises through the assignment of an identity number or unique name is critical for effective traceability of a product.

2. *Identification of product*: Every lot or batch of food material is given a unique identifier that accompanies it and is recorded at all stages of its progress through its food chain. It is easier to find and track products that have been assigned an identification number or code. Some examples are livestock identifiers (e.g., Radio Frequency Identification Device (RFID) ear tags), barcodes containing an assigned lot number, and harvest or production dates.

3. *Records of the movements*: Premises and product identification are prerequisites for movement recording. Movement of specific products from one location to another can be tracked. Records must be maintained that should have information on the identifiers of the product, quantity of product moved, date, shipper, transportation used, and receiver identification.

2.7.4 TECHNIQUES USED FOR FOOD TRACEABILITY

There are different types of equipment and techniques that are used for tracing and tracking a product through a supply chain. These range from paper-based records maintained by producers, processors, and suppliers to more sophisticated solutions. Commodity or market requirements decide the use of a particular technique. Product identification techniques and methods that can be used in the agriculture and food sectors are given as follows:

1. *Document-based techniques*: The simplest way of keeping track of products is by using pen and paper to record, store, and communicate data to partners in the supply chain. This is probably the oldest system, which is still equally reliable and cost-effective. Paper invoices, purchase orders, and bills of lading, as well as electronic file formats may be used to store alphanumeric

codes and other data on product lot number, harvest date, product receipt/ shipping date, quantity, or ingredients. However, searching through paper records is time-consuming and laborious.

2. *Structured database techniques*: Some organizations handle data in a structured manner and use management information systems and other databases such as enterprise resource planning systems for inventory control, warehouse management, accounting, and asset management. They can also prepare their own software as well. Structured databases enable rapid and precise queries of data elements to isolate the source and location of products that may be contaminated. Event system processing (ESP) systems such as Systems, Applications, Products in Data Processing (SAP) can read standardized data from barcodes and RFIDs, including Global Telecommunications Networks (GTNs) and Global Location Numbers (GLNs).

3. *Barcodes*: Barcodes are being used extensively these days. A typical barcode consists of a label that is applied to a product such as an individual item, box, or pallet. Barcode labels are cost-effective and can contain customized details about the product such as variety, size, or quantity. Barcodes require a printing component to print them on labels or products, and scanning technology to read barcoded information.

4. *Radio frequency identification devices (RFIDs)*: These are seen as an alternative to barcode systems. An RFID is a product identification device containing a small electronic chip and antenna that can transmit information wirelessly with an appropriate scanner. Information scanned and transmitted from an RFID tag can be in the form of an identification number or can include more data that serves to identify the product being scanned. Advantages of RFID tags include the ability to scan at long distances and scan multiple tags simultaneously, and that visual scanning is not required. RFIDs are used extensively in the livestock industry, as well as for pallet movement tracking by some retail operations.

5. *Nano techniques*: Nano solutions enable food safety and preservation. Nano materials may be used in smart packaging and in food handling to detect pathogens, gases, spoilage, and changing temperature and moisture. Current technologies to detect pathogens in food require considerable time, money, and effort. Nano solutions can detect contamination in real time (Karippacheri et al., 2011). Nanotechnologies are also enabling the production of cheaper and more efficient nano-scale RFIDs for tracking and monitoring food through the supply chain for traceability (Joseph and Morrison, 2006).

6. *DNA techniques*: While conventional methods of traceability work for labeling and tagging food products that are not genetically modified or engineered, DNA traceability offers a more precise form of traceability for animals and animal by-products derived through biotechnology. DNA traceability works on the principle that each animal is genetically unique and thus by-products of the animal can be traced to its source by identifying its DNA (Loftus, 2005). This is the key technology for species identification as there is no limitation when different processing treatments for fish and seafood are used.

7. *Nuclear techniques*: Nuclear techniques are tailored to determine the provenance of food by assessing its isotopic and elemental fingerprints. These techniques are also used to identify the geographical origin of food and to identify sources of contamination. These methodologies may be used for the traceability and authentication of food and have the potential to be applied in many developing countries, thereby enhancing their capacity to improve food safety and quality. Where appropriate, these techniques can also be complemented by conventional, non-nuclear approaches (International Atomic Energy Agency, IAEA).

8. *Foodomics*: Foodomics plays an important role in food traceability in which mass spectroscopy (MS)-based metabolomics and proteomics are applied. As an example, the profiling of metabolites can be used not only to determine the origin but also to obtain the traceability of a given food—that is, to precisely know all the different manufacturing steps to which a particular food has been submitted. Thus, powerful separation techniques such as comprehensive two-dimensional gas chromatography (2D GC) have been combined with high-resolution mass spectroscopy (MS) (time of flight-mass spectroscopy [TOF-MS]) to profile monoterpenoids in grapes (Rocha et al., 2007). This profiling has revealed the precise monoterpenoid composition of the different varieties of grapes, and its application has allowed for the traceability of the products directly derived from these grapes, such as musts and wines. A similar strategy of separation techniques was devised and aimed to the correct origin traceability of honey samples (Cajka et al., 2009).

The WHO Department of Food Safety and Zoonoses (FOS) provides scientific advice for organizations and the public on issues concerning the safety of food. Its mission is to lower the burden of foodborne disease, thereby strengthening the health security and sustainable development of member states. Foodborne and waterborne diarrheal diseases kill an estimated 2.2 million people annually, most of whom are children. The WHO works closely with the Food and Agriculture Organization of the United Nations (FAO) to address food safety issues along the entire food production chain—from production to consumption—using new methods of risk analysis. These methods provide efficient, science-based tools to improve food safety, thereby benefiting both public health and economic development.

2.7.5 INTERNATIONAL FOOD SAFETY AUTHORITIES NETWORK

The International Food Safety Authorities Network (INFOSAN) is a joint program of the WHO and the FAO. INFOSAN has been connecting national authorities from around the globe since 2004, with the goal of preventing the international spread of contaminated food and foodborne disease and strengthening food safety systems globally. This is done by

1. Promoting the rapid exchange of information during food safety events
2. Sharing information on important food safety issues of global interest
3. Promoting partnership and collaboration between countries
4. Helping countries strengthen their capacity to manage food safety risks

Membership of INFOSAN is voluntary but is restricted to representatives from national and regional government authorities and requires an official letter of designation. INFOSAN seeks to reflect the multidisciplinary nature of food safety and promote intersectoral collaboration by requesting the designation of focal points in each of the respective national authorities with a stake in food safety, and a single emergency contact point in the national authority with the responsibility for coordinating national food safety emergencies. Countries choosing to be members of INFOSAN are committed to sharing information between their respective food safety authorities and other INFOSAN members. The operational definition of a food safety authority includes those authorities involved in food policy, risk assessment, food control and management, food inspection services, foodborne disease surveillance and response, laboratory services for the monitoring and surveillance of foods and foodborne diseases, and food safety information, education, and communication across the farm-to-table continuum.

REFERENCES

CAC (Codex Alimentarius Commission). (2001). *Food Hygiene Basic Texts*. Rome, Italy: World Health Organization, FAO.
CAC (Codex Alimentarius Commission). (2003a). CAC/RCP 1-1969, Rev.4-2003. *Recommended International Code of Practice: General Principles of Food Hygiene*. Rome, Italy: World Health Organization, FAO.
CAC (Codex Alimentarius Commission). (2003b). A Hazard Analysis and Critical Control Point (HACCP) system and guidelines for its application [Annex to CAC/RCP 1-1969, Rev 4-2003]. Rome, Italy: World Health Organization, FAO.
CAC (Codex Alimentarius Commission). (2006). *Principles for Traceability/Product Tracing as a Tool within a Food Inspection and Certification System CAC/GL 60-2006*. Rome, Italy: World Health Organization, FAO.
CAC (Codex Alimentarius Commission). (2009). *Food Hygiene: Basic Texts*, 4th edn. Rome, Italy: World Health Organization, FAO.
Cajka T., Hajslova J., Pudil F., and Riddellova K. (2009). Traceability of honey origin based on volatiles pattern processing by artificial neural networks. *Journal of Chromatography A* 1216: 1458–1462.
European Commission. (2007). Fact sheet, food traceability, June, 2007. Directorate-General for Health and Consumer Protection European Commission: B-1049 Brussels Directorate-General for Health and Consumer Protection, European Commission: B-1049 Brussels. ec.europa.eu/food/safety/docs/gfl_req_guidance_rev_8_en.pdf. Accessed on October 27, 2015.
FAO/WHO. (2007). *FAO/WHO Guidance to Governments on the Application of HACCP in Small and/or Less-Developed Food Businesses*. Rome, Italy: World Health Organization, FAO.
IFT (Institute for Food Technologists). (2009). Traceability (product tracing) in food systems: An IFT report submitted to the FDA, volume 1: Technical aspects and recommendations/food and drug administration. *Comprehensive Reviews in Food Science and Food Safety* 9(1): 92–158. http://onlinelibrary.wiley.com/doi/10.1111/j.1541-4337.2009.00097.x/pdf. Accessed October 27, 2015.
International HACCP Alliance. (1999). Guidelines for Developing Good Manufacturing Practices (GMPs), Standard Operating Procedures (SOPs) and Environmental Sampling/Testing Recommendations (ESTRs) Ready-to-Eat (RTE) Products. http://haccpalliance.org/sub/food-safety/guifinal2.pdf, Accessed on October 27, 2015.

Joseph T. and Morrison M. (2006). *Nanotechnology in Agriculture and Food*. Report. Nanoforum. org European Nanotechnology Gateway. http://nanotech.law.asu.edu/Documents/2009/09/ nanotechnology_in_agriculture_and_food_234_2644.pdf, Accessed on October 4, 2015.

Karippacheril T.G., Rios L.D., and Srivastava L. (2011). Global markets, global challenges: Improving food safety and traceability while empowering smallholders through ICT. In *Module 12 in ICT in Agriculture Source Book*, The World Bank, Washington, D.C. http://www.ictinagriculture.org/sourcebook/module-12-improving-food-safety-and-traceability. Accessed on October 27, 2015.

Loftus R. (2005). Traceability of biotech derived animals: Application of DNA. *Scientific and Technical Review (OIE)* 24(1): 231–242.

Rocha S.M., Coehlo E., Zrostlikova J., Delgadillo I., and Coimbra M.A. (2007). Comprehensive two-dimensional gas chromatography with time-of-flight mass spectrometry of 62 mono-terpenoids as a powerful tool for grape origin traceability. *Journal of Chromatography A* 1161:292–299.

Shiklomanov I.A. (2000). Appraisal and assessment of world water resources. *Water International* 25(1): 11–32, International Water Resources Association.

Sperber, W.H. and Stier, R.F. (2010). Happy 50th birthday to HACCP: Retrospective and prospective. *Food Safety Magazine* December 2009–January 2010: 42, 44–46.

USDA-FSIS. (1999). *Guidebook for the Preparation of HACCP Plans*. (www.fsis.usda.gov/ index.htm).

USDA-FSIS. (2007). Code of Federal Regulations Title 9: Animals and Animal Production, Part 416 (www.access.gpo.gov/nara/cfr/waisidx_07/9cfr416_07.html).

WHO. (2010). Prevention of foodborne disease: Five keys to safer food. www.who.int/ foodsafety/areas_work/food-hygiene/5keysflyer2016.pdf?ua=1, Accessed on October 24, 2015.

3 Risks and Controls in the Food Chain

3.1 HISTORY OF FOOD SAFETY AND TRADITIONAL FOOD SAFETY SYSTEMS

The history of foodborne illness is as old as the human race. People must have been getting sick from food for as long as we have been eating it. The knowledge gained by earlier people must have come from eating a particular food in particular conditions; if it caused problems, then they must have tried to avoid the stale food or particular plants/fruits. The first suggested case of a known foodborne illness appears to have been that of Alexander the Great, who died in 323 BC. from a case of typhoid fever when he and his army stopped to rest in ancient Babylon. Typhoid fever is caused by the bacterium *Salmonella typhi*, which can be contracted from contaminated food or water. Although this theory can never be fully proven, it goes to show that humans have probably been affected by these illnesses throughout history. Other well-known people from history who are suspected to have died from foodborne illnesses include King Henry I, Rudyard Kipling, President Zachary Taylor, and Prince Albert. These types of incidents did help to create some kind of informal system to identify and avoid the food that can cause illness. So early on, along with the development of techniques for preserving foods, we also started adopting certain types of laws and regulations to control the handling of foods. Numerous "food taboos" were already known in ancient times. They often had a ritual and religious character, yet they always served food hygiene and safety. For instance, Judea prohibited the consumption of pork in 1800 BC and India developed a list of "unclean" foods in 500 BC; this list included meat cut with a sword, dog meat, human meat, and much more. The Middle Ages were another period in which food regulations continued to be developed. King John of England passed the first food law in 1202. In the United States, foodborne pathogens have played roles in settling territory and fighting wars. Many historians believe that the first English settlement in Jamestown, VA, was decimated by typhoid fever many times between 1607 and 1699, ultimately leading to its demise. Also in the late 1600s, a toxic fungus changed the course of history and led to the Salem witch trials. The fungus, which was growing on the rye they used for food, caused many symptoms with which settlers were unfamiliar, which led to the accusations of witchcraft and the killing of those infected. In 1898, typhoid fever struck again during the Spanish-American war, infecting more than 20,000 American soldiers.

Food safety has been an issue being handled over the centuries in different manners. Many of the toxins and spoiled food were proven to have harmful effects on human health through trials and experiences. Thus, in all traditional societies, spoiled food and some parts of plants and animals were not consumed. We see that a few types of mushrooms, green vegetables, and some portions of certain animals were not eaten as the tribal populations had good practical knowledge systems developed about this. It was also known that food kept for a long time becomes unsafe to consume. Over the centuries, a lot of research was also conducted to establish the reasons for such spoilage and find possible solutions. However, the system, which was working and still holds true, is based on the fact that the presence of a hazard in a food makes it unsafe. Over the course of the past 20 years, we have seen the emergence of risk analysis as the foundation for developing food safety systems and policies. This period has witnessed a gradual shift from a "hazard-based approach" in food safety to a "risk-based approach." A risk-based approach is basically a determination as to whether the exposure to a hazard has a meaningful impact on public health. It is now common to see industry, governments, and consumers all call for the adoption of risk-based programs to manage chemical, microbiological, and physical risks associated with the production, processing, distribution, marketing, and consumption of foods. However, there is not always a clear understanding about what risk management encompasses or about the consequences of developing risk management systems. Accordingly, it is worth exploring some of the food safety risk management concepts and principles that are emerging both nationally and internationally and that are being used to establish systems that are likely to dramatically transform the food safety landscape in the next 10 years.

The use of risk management concepts is not new. In fact, many of the much-discussed differences between hazard- and risk-based approaches disappear when examined more closely. If one carefully examines most traditional hazard-based food safety systems, at their core are one or more key risk-related assumptions that drive the decision-making process. These risk-based decisions may not be made consciously or may not be so scientifically thought of, but they nevertheless effectively introduce decision criteria related to relative risk. The most common is the establishment of standard methods of analysis for verifying that a food system is controlling a specific hazard. Through the selection of the analytical method, the sampling plan, and the frequency of testing, a risk-based decision (i.e., a decision based on the frequency and severity of the adverse effects of a hazard) is introduced. Typically, if a hazard occurs frequently and/or has a severe consequence, the frequency and sensitivity of analytical testing are made more stringent than if the hazard is infrequent and/or has mild adverse effects. The practice of stringent analytical testing in cases of frequent hazard was being followed. Based on experience, we were forced to systematize and establish standard protocol so that the same mistakes were not made each time. For example, traditional sampling plans for powdered infant formula generally called for testing each lot for *Salmonella enterica* at a level of absence in 1.5 kg (sixty 25 g samples) versus the sampling of refrigerated, ready-to-eat foods for *Staphylococcus aureus*, which is done only infrequently and typically allows the microorganism to be present at levels between 100 and 1000 CFU/g. Obviously, while these two microorganisms are both food safety hazards, risk-based decisions are made related to the stringency needed to control the hazard. This is dependent on the degree and frequency of the potential harm.

Thus the actual difference between hazard- and risk-based systems is only to do with how uncertainty and transparency are dealt with during the decision-making process so that food is safe (Ackerley et al., 2010; Batz et al., 2011).

3.2 FOOD SAFETY AND GLOBALIZATION OF THE FOOD CHAIN

There has been a major change in the supply chain all over the globe, especially due to increased communications and trade. The products that are produced in one country are processed in another and subsequently marketed in a third country, so food products and articles are constantly moving from one country to another. During the last few decades, in almost all countries, there have been changes in food laws, regulations, and standards, mainly in order to address the concerns of the consumers. These have put the responsibility on the food industry to source, produce, and deliver safe and high-quality food. Therefore, the rising costs associated with quality and safety, along with traceability and recall issues, have been impacting the food chain. It may be true that the number of prosecutions and recalls is increasing, but at the same time overall food safety is improving. In 2014, the U.S. Food and Drug Administration (FDA) registered 253 food recalls in the first two quarters of the year alone. All over the world, and especially in Europe, there were hundreds of recalls, withdrawals, and safety alerts throughout the year. Even in India, in 2015, the recall of Nestlé's Maggi noodles made the news and created significant controversy. In Asian countries, the trend toward recall and safety alerts is also on the increase. The disadvantages associated with these controversies are that they are sometimes blown out of proportion, which creates doubt about the vulnerability of the food supply chain. Consumers get confused and panic spreads in the market. Misinformation and misconceptions spread very quickly. However, these controversies also do some good, as we start looking at the loopholes in the supply chain and making necessary amendments and improvements. At the same time, the food industry is under tremendous pressure to regain consumer trust by delivering safe and high-quality food while meeting regulatory requirements across its complex supply chain (Buchanan, 2010). The food supply chain is becoming more complex day by day, mainly due to the movements of product over long distances and also because of the length, processes, and duration of the supply chain. Each and every operation in this chain is unique and offers ample opportunities for the deterioration of food in quality as well as safety. Thus the food organizations have to create adequate initiatives to identify and assess risks, take precautionary measures to control potential food safety incidents, and comply with the regulatory mandates. There is a need to have a risk-based preventive approach in which data and expertise are compiled and analyzed so as to identify potential food safety issues. This has to be done at each stage along the production, distribution, and handling chain. There is a need to have a proactive approach to handling the issue of food safety and quality. This will also help to establish the confidence of the consumers.

It is not that the globalization of food supply chains has made them more susceptible to lapses in quality. In fact, the competition over and awareness of nutrition has ensured that quality and safety is better. Suppliers are reaching into all corners of the world to satisfy consumer demand for year-round availability of fresh produce. Along with this, the improvements in regulations and food control systems, new standards, and the availability of improved technology has resulted in overall improvement of systems in food safety. The globalization of trade has also resulted in a demand for improving food control systems. There have been regulation changes and also voluntary actions by companies to improve the image of their brands. India came out with a comprehensive Food Act in 2006. Most of the Asian countries have also either come up with new regulations or improved earlier existing ones. The European Union has also come out with new regulations. In 2011, the United States felt a need to have new a law—the Food Safety Modernization Act (FSMA)—that mandates a series of strict "prevention-based" controls across the supply chain. The commonality among all this new legislation is that it emphasizes prevention and risk-based actions. These new regulations stress self-regulation by food business operators, as it has become evident that self-regulation is the only way to ensure the safety of food at each level of operations. Food business operators are being called on to do a better job of tracing the origins of their products, right down to the individual ingredients. It is recognized that it is not practically possible to check each and every product for safety because the volume is so high, but each and every food article needs to be safe, so risk-based studies are done and strategies are formulated accordingly to take samples, do tests, and conduct the right actions at appropriate levels in the food chain.

For their part, consumers are demanding more information about food items that claim to be novel foods, functional foods, health foods, nutraceuticals, and organic foods. Many of these products haven't been shown to pose a health or safety issue, but consumers are definitely concerned about the accuracy of the claims made. The same level of sensitivity applies to the use by food manufacturers of the so-called proprietary label, which they are under increasing pressure to justify. The industry has responded to consumer concerns by creating a number of safety and quality schemes. Through these common platforms, the industry wishes to reach across global supply chains. The full implications of these initiatives by industry and governments have yet to be felt. It has been established through these initiatives that industry must adopt a "food safety culture" that promotes a deep understanding of health and safety issues. There should not be any attempts to cut corners on health, safety, and quality controls. Food producers and retailers need to tackle the issues of risk and quality head on.

3.3   FOOD SAFETY CHALLENGES IN THE GLOBAL FOOD SAFETY CHAIN

It has been discussed in Chapter 1 that food safety has been identified as a public health priority at the international level, as unsafe food causes illness and death in millions of people every year. According to the World Health Organization (WHO, 2010), there have been serious incidences of outbreaks of foodborne diseases all over the world. Efforts are being made at all levels to minimize health risks from farm to table, to prevent outbreaks, and to promote food safety. Governments are taking keen interest in this due to the enormous economic impact of diseases and health concerns on the population. These incidents of outbreak have an adverse impact on trade and economic development. There is a loss of productive life as well as productive

FIGURE 3.1   Food safety challenges in the global food safety chain.

man-days. Spending on medical care is to be increased. A few of the major global food safety concerns in the food chain, as given in Figure 3.1, are as follows:

1. *Food security challenges*: It has been discussed in previous chapters that food security exists when everyone has access to safe, nutritious, affordable food, which is the foundation for an active and healthy life. As the world's population growth increases, the issues of food accessibility and availability are becoming major. By 2050 the world's population is expected to reach 9 billion—an increase of 2 billion people from the present level—which will put a strain on already scarce resources like land and water. In the process of maintaining food production levels and providing food to everyone, there are concerns that food safety, food quality, and environmental balance should not be compromised. The race to produce more food may bring in complexities in seed production and crop protection, thereby making food safety more complex.

2. *Challenges within the food supply chain*: Spoilage of food through the spread of microbiological hazards and chemical food contaminants, adulteration (intentional or unintentional), quality and consistency, availability and seasonality, and the cost of keeping agriculture remunerative and food affordable are major challenges at all levels. Since these are issues that require detailed appreciation, they have been discussed under the heading "challenges between food supply chains" further in this chapter.

3. *Assessments of new food technologies and innovations*: With more consumer awareness of and interest in health improvement, innovations are being carried out and new technologies are being used. Novel foods, genetically

modified foods, and foods sold for therapeutic purposes are some of the results of such aspirations. These are to be tested for safety and should also not have any adverse impact on the environment.

4. *Regulatory challenges*: All these inventions and innovations in food and the globalization of the food trade have also impacted on the regulatory systems of countries. Standards are to be upgraded and analytical processes are to be strengthened on a regular basis. States have been responding and making changes to their regulations so that a strong food safety system is in place to ensure a safe global food chain. Some of the countries that are not able to cope with changes face unique regulatory challenges to bring themselves up to acceptable food safety standards.

5. *Increased global conflict and tension*: This has a multifarious impact. Food security becomes a major issue, but diversion of economic and other resources to war-related expenditure takes its toll on the health and well-being of the people.

6. *Global recession and financial market disruption*: This also has a direct impact on food security and poverty, and food safety is also compromised.

These global issues require a multidisciplinary and collaborative approach to develop responses effective at domestic, regional, and international levels.

3.4 CHALLENGES WITHIN THE FOOD SUPPLY CHAIN

It is no secret that in recent years there has been significant media exposure of some food safety incidents, with a notable public health impact and decrease in consumer confidence. This has increased the focus on food safety on the part of consumers, the industry, lawmakers, and regulatory agencies. Consumers have become more knowledgeable and demanding and expect that the system should take steps to address the top food safety challenges facing the industry. The one overriding food industry concern is how to best manage food safety across the global supply chain while ensuring regulatory compliance. There is a need to establish a system that identifies the action steps that can be taken to mitigate the inherent risks of a global food supply chain. There have been some positive outcomes of this awareness about food safety. The increased awareness of some of these large-scale foodborne illness and contamination events has resulted in advances and improvements in public health signal detection. There have been new regulatory reporting requirements of contaminated food products in commerce. This has also resulted in improved communication streams and interconnectivity between regulatory agencies, domestically and internationally. Steps have been taken to create a system whereby state public health laboratories can analyze strains of certain pathogenic bacteria from ill individuals and determine their genetic fingerprint. This information is then shared nationally with the Center for Disease Control and Prevention. Surveillance exercises are conducted to evaluate whether there is a common food or environmental source of the genetically matched strain that made people ill. The upstream and downstream distribution supply chain partners may need to submit their distribution and handling information, because most of the time the origin of a hazard or problem lies in the supply chain.

An ever-increasing complex global food and ingredient supply has introduced further opportunities for contamination to be incorporated into food products. To overcome the threat of contamination in the supply chain, we need to understand the genesis, type, and cause of these contaminations and spoilages. The spoilages can be of different types and origins; spoilage is the most important challenge, and there are also five other major challenges in the food supply chain.

1. *Spoilage*: Food spoilage can be defined as a disagreeable change in a food's normal state. These changes may make food unsafe or unacceptable due to changes in odor or appearance. Such changes can be detected by smell, taste, touch, or sight. The food chain faces four main types of change:

   a. *Microbial*: There are three types of microorganisms that cause food spoilage: yeasts, molds, and bacteria. Both yeasts and molds are fungi. However, mold is a type of fungus that grows in multicellular filaments called "hyphae." These tubular branches have multiple, genetically identical nuclei, yet form a single organism known as a "colony." In contrast, yeast is a type of fungus that grows as a single cell. Yeast growth causes fermentation, which is the result of yeast metabolism. Molds grow in filaments, forming a tough mass that is visible as "mold growth." Molds form spores that, when dry, float through the air to find suitable conditions in which they can start the growth cycle again. Both yeasts and molds can thrive in high-acid foods like fruits, tomatoes, jams, jellies, and pickles. Both are easily destroyed by heat. Processing high-acid foods at a temperature of 212°F (100°C) in boiling water for the appropriate length of time destroys yeasts and molds. Bacteria are round, rod-, or spiral-shaped microorganisms. Bacteria may grow under a wide variety of conditions. There are many types of bacteria that cause spoilage. They can be divided into spore-forming and non-spore-forming bacteria. Bacteria generally prefer low-acid foods like vegetables and meat. In order to destroy bacteria spores in a relatively short period of time, low-acid foods must be processed for the appropriate length of time at 240°F (116°C). Eating food spoiled by bacteria can cause food poisoning.

   b. *Autolysis*: More commonly known as self-digestion, this refers to the destruction of a cell through the action of its own enzymes. It may also refer to the digestion of an enzyme by another molecule of the same enzyme. Enzymes are proteins found in all plants and animals. If uncooked foods are not used while fresh, enzymes cause undesirable changes in color, texture, and flavor. Enzymes are destroyed easily by heat processing.

   c. *Oxidation by air*: Atmospheric oxygen can react with some food components, which may cause rancidity or color changes. Oxidative rancidity is a major cause of food quality deterioration, leading to the formation of undesirable off-flavors as well as unhealthful compounds. Antioxidants are widely employed to inhibit oxidation, and with current consumer concerns about synthetic additives and natural antioxidants these are of much interest.

d. *Other factors*: Infestations (i.e., invasions) by insects and rodents also account for huge losses in food, especially when stored. Low-temperature injury also destroys food as cells are injured and harmed; the internal structures of the food are damaged by very low temperatures (e.g., the internal mahogany browning of a potato or chilling injuries in cucumbers are low-temperature injuries). Sometimes this also induces microbial spoilage. All these changes in the food are caused by a number of things: air and oxygen, moisture, light, microbial growth, and temperature.

   i. *Air*: One of the major reasons for food spoilage is air. Air consists of 78% nitrogen, 21% oxygen, and a 1% mixture of other gases. While oxygen is essential for life, it can have deteriorative effects on fats, food colors, vitamins, flavors, and other food constituents. Oxygen can cause food spoilage in several ways. It can: provide the conditions that will enhance the growth of microorganisms; cause damage to foods with the help of enzymes; and cause oxidation. Oxidative spoilage is the chief cause of quality loss in fats and fatty portions of foods. When lipids oxidize, short-chain carbon compounds are formed; these compounds have very strong odors and flavors and are very undesirable and unacceptable. The off-odors resulting from this type of spoilage are sharp and acrid and have been described as linseed-oil-like, tallowy, fishy, or perfume-like.

   ii. *Microorganisms*: Oxygen and moisture can provide conditions that enhance the growth of microorganisms. Some bacteria require oxygen for growth ("aerobes"), while others can grow only in the absence of oxygen ("anaerobes"). Many bacteria can grow under either condition and are called "facultative anaerobes." Molds and most yeasts that cause food to spoil require oxygen to grow. They can often be found growing on the surface of foods when air is present.

   iii. *Enzymes*: Certain enzymes that are naturally present in foods are known as "oxidizing enzymes." These enzymes catalyze chemical reactions between oxygen and food components, and this leads to food spoilage. Although there are many oxidizing enzymes, two that can cause darkening in diced and sliced vegetables are catalase and peroxidase. The browning of vegetables caused by these enzymes is often accompanied by off-flavors and odors. A simple heat treatment (e.g., blanching) is used to inactivate these enzymes.

   iv. *Water*: Water is one of the most common substances on earth. It is an essential component of all foods, and all living organisms also contain water. The amount of water in a food (known as "percent water") influences the appearance, texture, and flavor of the food. Water makes up about 70% or more of the weight of most fresh (i.e., unprocessed) foods. Even "dry" foods like beans, flour, and cereals contain some water. Fresh fruits and vegetables contain the most water—between 90% and 95% water. Free and available water in any food is the main cause for inducing spoilage.

2. *Adulteration*: Within the supply chain, adulteration is the key food safety issue. Adulterated food is that which is generally impure, unsafe, or unwholesome. Adulteration can be unintentional or intentional and can cause biological, chemical, or physical hazards. Intentional adulteration can be for economic or other reasons.

3. *Quality and consistency*: Consistency is key in the food industry. A food product's recipe must continually be accurate and repeatable, not only to maintain high quality but also to ensure safety. Quality and consistency are the cornerstones of consumer trust in a company's food products and brand. From helping suppliers develop credible certificates of analysis to conducting internal quality analysis of raw, in-process, and finished products, companies needs to capture the quality data to create consistency.

4. *Availability and seasonality*: Seasonality of food refers to the times of year when a given type of food is at its peak, either in terms of harvest or flavor. This is usually the time when the item is cheapest and freshest on the market. The food's peak time in terms of harvest usually coincides with when its flavor is at its best. Availability and seasonality of food offers a challenge to the supply chain.

5. *Prices*: The interest in marketing margins and price transmission has recently gained remarkable momentum and the number of studies on this subject is rapidly growing. There are a number of questions to be answered about the prices and margins being charged, yet new questions are still surfacing as markets and business practices keep changing. There are demands that farmers invest in more sustainable food production systems, sensitive to concerns about the environment and the use of scarce national resources. These demands cannot be met by putting relentless pressure on producer prices and their margins, which are decreasing. The powerful in the food chain are imposing conditions on the powerless in an unfair manner, leading to reduced margins, impossible specification demand, and poor returns for investment outlay.

All the challenges faced within a typical food chain have been summarized in Figure 3.2.

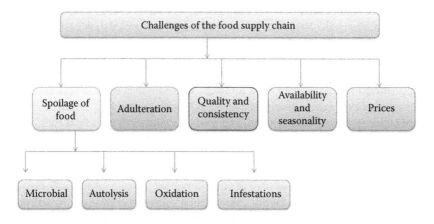

FIGURE 3.2   Challenges within the food safety chain.

Global food supply continues to grow in volume and complexity (Buzby et al., 2008). Imports are expected to continue to grow because of cost concerns (i.e., the need for lower costs and higher productivity), availability (including seasonality), and consumer demand for diverse food products. According to an FDA (2011) report entitled "Pathway to Global Safety and Quality," between 10% and 15% of all food consumed in the United States is imported. According to the U.S. Government Accountability Office, imports account for nearly two-thirds of the fruits and vegetables and 80% of the seafood eaten domestically. In the United States and also most of the countries in Europe, seafood and spices are among the most imported food items. One of the reasons for this is that consumers are now growing accustomed to having a wide selection of food products from around the world. There has been an increase in food trade and the growth of trade in food products from countries like Mexico, China, Asia, India, and Africa. According to an analysis of the food products refused by the FDA at one of the ports, conducted by the U.S. Department of Agriculture (USDA) Economic Research Service, "The three food industry groups with the most violations were vegetables (20.6 % of total violations), fishery and seafood (20.1%) and fruits (11.7%). Violations observed over the entire time period include sanitary issues in seafood and fruit products, pesticides in vegetables and unregistered processes for canned food products in all three industries."

Since the United States has compiled available data, analysis of the food trade has become easier (Gary et al., 2011). The U.S. data shows that food products are imported from more than 300,000 facilities in 150 countries, so we can have idea of the complexity involved. The FDA has a system for inspecting the facilities that send food to the United States, and such facilities are to be preapproved. This definitely enhances the workload but ensures compliance as well. Trade in food will continue to grow globally because of the rise of emerging markets, the scarcity of natural resources, and the increasing flow of capital, information, and goods across borders. With this increase will come increased complexity for regulators, as the distinction between foreign and domestic products continues to blur and becomes ever more complex. With this complexity comes the challenge of being able to trace the products, and to ensure that all entities within the supply chain meet their responsibilities for food safety and quality.

Supply chain management is gaining importance as incidences of poor-quality food products show that a sound control strategy is required and risk analysis is very important. Chen et al. (2014) studied the managerial and policy issues related to quality control in food supply chain management with a focus on the Chinese dairy industry. Based on a general supply chain model with acceptance sampling tests under uncertain product quality, they showed that, depending on the sampling technology, the decentralized supply chain structure may lead to a distortion in product quality. They used an exploratory case study of the 2008 adulterated milk incident in China to investigate practical issues in ensuring product quality/safety in food supply chain management. They concluded that, instead of the common "poor quality" misperception of food products from global emerging markets, it was actually the poor vertical control strategy for managing the food supply chain's quality and risk that caused the adulterated milk incident. A number of other important managerial and policy insights and implications regarding supply chain design, informational visibility, corporate social responsibility, and regulatory action in managing the global food supply chain's quality and risk were also found to be relevant (Nepusz et al., 2009).

3.5   SCIENCE-BASED APPROACH TO FOOD SAFETY

It is clear that, due to complexity and increased volume, maintaining the safety and integrity of food is not easy. It involves the collection of information related to health and illness and the impact of chemicals, laboratory analysis, the creation of models based on these results and information, and the management and communication of these models and strategies. This all is possible only through systematic and scientific study and analysis. It is difficult to conceive of a food safety system that responds effectively and efficiently to emerging food safety concerns, which does not permit rapid changes in approach based on advances in science. The collection of data itself is a huge exercise, and sampling requires a system to be followed so as to make it truly representative. New hazards are identified due to advancements in agricultural and production techniques. The flexibility to respond to new information and hazards will require unfettered data sharing. Data analysis should lead to control measures, and the acceptance of any process is possible if a system emphasizes validation and verification of the control methods used. The scientific methodology is used in the following processes for assuring food safety:

1. *Surveillance systems*: Our understanding of the epidemiology and sources of foodborne diseases can be improved through the scientific study of human, animal, and environmental surveillance systems. Human foodborne disease surveillance is a very important aspect to knowing the impact of food on the spread of disease. Surveillance helps to identify outbreaks of foodborne diseases so they can be controlled and prevented. It can also determine the causes of a foodborne disease. This results in monitoring the spread of disease so that control strategies can be improved.
2. *Risk assessment*: Risk assessment is a prerequisite for selecting food safety management options. Risk assessment is a mathematical computational process that is done scientifically to collect, collate, and update information on risk. Although essential, scientific data is a very substantial limiting factor in the application of risk assessment. Appropriate and aggressive data collection throughout the food production and processing system is essential for valid risk assessments and the resulting food safety improvements. Systems are to be prepared so that data can be properly evaluated and the results made available.
3. *Risk management*: Successful management of risk is the main objective of any food control system. Food regulatory agencies work with public health systems, industry, other food business operators, and consumers to establish Food Safety Objectives (FSOs). FSOs offer a means to convert public health goals into values or targets that can be used by regulatory agencies and food manufacturers. FSOs, which can be applied throughout the food chain, specify the maximum level of hazard that would be appropriate at the time a food is consumed. FSOs enable food business operators to design processes that provide the appropriate level of control and that can be monitored to verify effectiveness. The FSO approach can be used to integrate risk assessment and current hazard management practices into a framework that achieves public health goals in a science-based manner.

4. *Good agriculture and manufacturing practices*: Good, well-defined, science-based agricultural practices should be further developed for specific commodities. Additional research will be necessary to better understand the microbial ecology in these agricultural environments and to formulate science-based recommendations for pathogen control. Similarly, good manufacturing practices are very important for monitoring the manufacturing process. Routine microbiological testing is useful for achieving and designing good agricultural and manufacturing practices. It can focus on pathogens of interest or on nonpathogenic microorganisms whose presence indicates conditions favorable to the presence of pathogens.

5. *Hazard analysis, control, and monitoring*: Hazard Analysis and Critical Control Points (HACCP) is a science-based food safety management approach that has been widely adopted and effectively applied to improve food safety. One can prepare a valid HACCP plan only after scientific analysis has identified any point that meets the critical control point (CCP) criteria. HACCP implementation must incorporate scientific knowledge and data in a product- and process-specific manner that best meets the FSOs. Testing is useful for surveillance and HACCP verification purposes. It is also used for validating and revalidating control procedures.

6. *Quality testing*: Quality testing, which mainly involves microbiological testing of finished products, is done to ensure safety. Testing has statistical limitations based on the amount of product sampled, the percentage of product that is contaminated, and the uniformity of the distribution of contamination throughout the food. As the amount of contamination in the food decreases, the food safety emphasis should focus on further controlling processing conditions through the application of science-based HACCP systems.

All the previously mentioned six points, as depicted in Figure 3.3, are the main pillars of the science-based approach to food safety.

Managing food safety risk is like playing a game of probability: one may not be always sure, but probability can be predicted based on certain available information. Food business operators must understand the "odds" of producing a safe product. They need to understand the odds that their purchasing, handling, and preparation practices could lead them to introduce a food safety hazard, and then modify their practices to minimize the chances of creating unsafe food and improve their odds of creating safe food. This is the advantage that risk management approaches bring to food safety: they increase the amount of knowledge mobilized to make better food safety decisions. It has to be understood that, if probability has been used as a tool, it does not mean that there is an element of uncertainty and that safety may be compromised. There is involvement of the scientific approach in taking samples, analyzing these, identifying hazards, and making probability models to arrive at appropriate decisions. Adoption of the risk management approach does not mean that there is a decrease in the level of protection afforded to the consumer. In fact, there will be a need for understanding scientific concepts, describing the concepts of variability and uncertainty, and explaining why food safety decisions cannot be based on "zero risk."

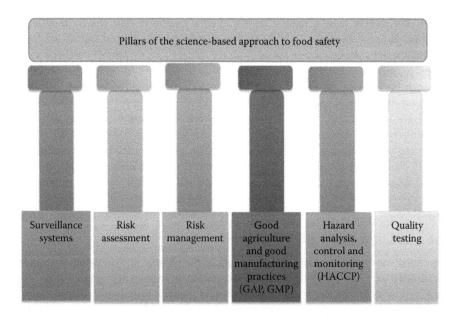

FIGURE 3.3 Pillars of the science-based approach to food safety.

When we calculate the safe limits of pesticides in food, we determine the toxicity of a compound in one or more animal models to identify the values at which it is safe. The safe levels are calculated and these are estimated against the likely-per-day consumption of food containing that compound, and safe values are projected for humans. Detailed experimentation takes place before we arrive at any decision about fixing a safe dose of that compound. Finally, the result arrived at is the result of a scientifically researched decision that weighs population susceptibility, likely levels of exposure, and the status of our knowledge about the toxicant's mechanism of action.

Convertino and Liang (2014) expressed the idea that scalable models that address the need to integrate epidemiological, social, and trade information into an operation research setting are in strong demand to reduce the U.S. and global public health risk. They proposed a model for the assessment of the potential health risk of food commodities based on the food supply chain as a subset of the international agro-food trade network. The model integrates concepts of network science in the supply chain and risk factors related to the food life cycle that occurs along the food supply chain (FCS). They consider the food pathogen risks from production to distribution, screening and manufacturing flaw risks, transportation and intermediary country risks, and country and manufacturer risks. Considering the safety of each country and the network variables, they introduced a global safety index (GSI) for characterizing the riskiness of each country based on local and food supply chain variables. The potential health risk is characterized by a multimodal distribution, and a ranking of food-pathogen-country triples reveals the unsafe paths of the food supply chain. Global sensitivity and uncertainty analyses show that network variables are driving the potential health risks, thus understanding them is crucial for public health risk management. The intermediary country risk, the food pathogen health risk, and company reliability are the second most important factors for potential health risks. Policies that act on both the supply chain variables and the safety index by means of the GSI reduce the average health risk by 44%. This reduction is much larger than the reduction of policies focused on individual risk factors of the food life cycle. Complex foods composed of multiple ingredients are among the riskiest foods, and their risk is driven by the food supply chain's complexity. Current management practices are focused on intervention after local outbreak cases and food pathogen relationships, with little attention paid to prior global food supply chain interventions. The FSC model here presented is claimed to be scalable to any level of the global food system and offers a different perspective upon which global public health is conceived, monitored, and regulated.

3.6   FOOD SAFETY RISK ANALYSIS

Risk analysis is used to develop an estimate of the risks to human health and safety, to identify and implement appropriate measures to control the risks, and to communicate with stakeholders about the risks and measures applied. It can be used to support and improve the development of standards, as well as to address food safety issues that result from emerging hazards or breakdowns in food control systems. It provides food safety regulators with the information and evidence they need for effective decision making, contributing to better food safety outcomes and improvements in public health. Regardless of the institutional context, the discipline of risk analysis offers a tool that all food safety authorities can use to make significant gains in food safety. For instance, risk analysis can be used to obtain information and evidence on the level of risk of a certain contaminant in the food supply, helping governments to decide which, if any, actions should be taken in response (e.g., setting or revising a maximum limit for that contaminant, increasing testing frequency, reviewing labeling requirements, providing advice to a specific population subgroup, issuing a product recall, and/or banning imports of the product in question). Furthermore, the process of conducting a risk analysis enables authorities to identify the various points of control along the food chain at which measures could be applied, to weigh up the costs and benefits of these different options, and to determine the most effective one(s). As such, it offers a framework for considering the likely impact of the possible measures (including on particular groups such as a food industry subsector) and contributes toward the enhanced utilization of public resources by focusing on the highest food safety risks. Risk analysis is comprised of three components: risk management, risk assessment, and risk communication. Each of these components has been applied in almost all countries for a long time, even before they came to be called by these names (see Figure 3.4). During the past two decades or so, the three components have been formalized, refined, and integrated into a unified discipline, developed at both the national and international levels, and are now known as "risk analysis." Risk analysis is the systematic study of the uncertainties and risks we encounter in business, engineering, public policy, and many other areas. Risk analysts seek to identify the risks faced by an institution or business unit, to understand how and when they arise, and to estimate the

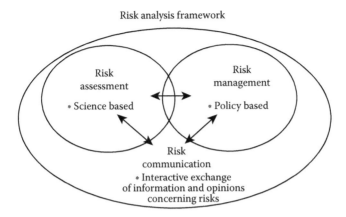

Risk analysis framework

FIGURE 3.4   Risk analysis framework.

impact (financial or otherwise) of adverse outcomes. Risk managers start with risk analysis, and then seek to take actions that will mitigate or hedge these risks. Risk analysis can be qualitative or quantitative. Qualitative risk analysis uses words or colors to identify and evaluate risks, or presents a written description of the risk, while quantitative risk analysis calculates the numerical probabilities of the possible consequences. Quantitative risk analysis is the practice of creating a mathematical model of a project or process that explicitly includes uncertain parameters that we cannot control, and also "decision variables" or parameters that we can control (Sinha and Lopez, 2014).

Based on similar lines, food safety risk analysis is essential for producing or manufacturing high-quality goods and products to ensure safety and protect public health. It should also comply with international and national standards and market regulations. Food safety risk analyses focus on major safety concerns in production processes. Not every safety issue requires a formal risk analysis (Food and Agriculture Organization, 2006).

Figure 3.4 illustrates the relationship between the three components of risk analysis.

The three main components of risk analysis have been defined by the Codex Alimentarius Commission (CAC) as follows:

1. *Risk assessment*: A scientifically based process consisting of the following steps:
   a. Hazard identification
   b. Hazard characterization
   c. Exposure assessment
   d. Risk characterization
2. *Risk management*: The process, distinct from risk assessment, of weighing policy alternatives in consultation with all interested parties, considering risk assessment and other factors relevant to the health protection of consumers and to the promotion of fair trade practices, and, if needed, selecting appropriate prevention and control options.

3. *Risk communication*: The interactive exchange of information and opinions throughout the risk analysis process concerning risk, risk-related factors, and risk perceptions among risk assessors, risk managers, consumers, industry, the academic community, and other interested parties, including the explanation of risk assessment findings and the basis of risk management decisions.

Risk assessment is considered to be the "science-based" component of risk analysis, while risk management is the component in which scientific information and other factors—such as economic, social, cultural, and ethical considerations—are integrated and weighed in choosing the preferred risk management options. In fact, risk assessment may also involve judgments and choices that are not entirely scientific, and risk managers need a sound understanding of the scientific approaches used by risk assessors. The interactions and overlaps of science and nonscientific values at various stages in risk analysis will be explored in more detail in subsequent chapters concerned with risk management and risk assessment (FAO, 2006).

Use of risk analysis: Risk analysis has been accepted as a potent tool for achieving food safety. The globalization of trade and the demands of consumers have created pressure for the establishment of a reliable, scientific system to ensure quality, and thus governments and regulators are making efforts to adopt appropriate control measures. Maintaining the safety of food requires constant vigilance by governments, industry, and consumers as the food supply changes as a result of new technologies, expanding trade opportunities, ethnic diversity in the population, and changing diets. Food regulators aim to ensure health and safety risks from food are negligible for the whole population, and that consumers can make informed choices. When this is achieved, public confidence in the effectiveness of food regulation is maintained. Food regulators develop and review food standards covering the composition and labeling of food sold, and address food safety issues and primary production and processing. They use risk analysis—an internationally accepted process—for standards development and many other situations where food-related health risks need to be assessed, managed, and communicated. Regulators are using a risk analysis framework for the following:

1. *Development of food standards*: The study of risk analysis has been always used in developing specifications and standards, whether vertical standards (which apply to a particular industry or to particular operations, practices, conditions, processes, means, methods, equipment or installations) and horizontal standards (other [more general] standards applicable to multiple industries). This is a very useful tool in developing new food standards.
2. *Evaluation and review of existing standards*: Risk analysis studies are also used for the evaluation of existing standards and for suggesting changes as substantiated through the studies.
3. *Surveillance for safe food*: Surveillance and monitoring plans are based on risk analysis studies. A food safety plan is designed for any establishment based on the risk profile, and thus the monitoring and surveillance activities are planned accordingly.

4. *Assessment of food processing practices*: The efficacy and efficiency of processing techniques are evaluated through risk analysis. There is a requirement to assess food technology practices to ensure that the product is safe and as per desired standards.
5. *Addressing the food safety concerns of consumers*: The food safety concerns of consumers are addressed through risk analysis and risk management, which are able to create an atmosphere of confidence that is very relevant, especially considering emerging food safety issues raised by consumers.
6. *Aid in regulatory decision making*: Using the risk analysis framework is intended to ensure effective regulatory decisions that can be justified and substantiated.
7. *Communication on food safety*: Its use encourages communication between all stakeholders, including consumers.
8. *Management of health risks*: Risk analysis in food regulation provides a basis for how regulators can use the risk analysis framework to manage a diverse range of food-related health risks.

Risk analysis must occur in a context, and to be done effectively it requires a formal process. In a typical instance, a food safety problem or issue is identified and risk managers initiate a risk management process, which they then see through to completion. This is best accomplished within a systematic, consistent, and readily understood framework in which scientific knowledge of risk and the evaluation of other factors relevant to public health protection are used to select and implement appropriate control measures. The responsibilities of risk managers during this process also include commissioning a risk assessment when one is needed, and making sure that risk communication occurs wherever necessary.

3.6.1 Risk Assessment

Risk assessment is the central scientific component of risk analysis and has evolved primarily because of the need to make decisions to protect health in the face of scientific uncertainty. Risk assessment can be generally described as characterizing the potential adverse effects to life and health resulting from exposure to hazards over a specified time period. Risk management and risk assessment are separate but closely linked activities, and ongoing, effective communication between those carrying out the separate functions is essential. Risk managers applying the Risk Management Framework (RMF) must decide whether a risk assessment is possible and necessary. If this decision is affirmative, risk managers commission and manage the risk assessment, carrying out tasks such as describing the purpose of the risk assessment and the food safety questions to be answered, establishing risk assessment policy, setting time schedules, and providing the resources necessary to carry out the work. While the main focus is on the application of risk assessment methodology as defined by the CAC, a broader view of risk assessment is also taken. All methods for assessing risk described here use the best scientific knowledge available to support risk-based standards or other risk management options.

Risk assessments incorporate, in one way or another, the four analytical steps described by the CAC. The way these steps are applied for microbiological and chemical hazards may differ in the details, but the basic structure remains the same. These four steps are

1. *Hazard identification*: The identification of biological, chemical, and physical agents capable of causing adverse health effects that may be present in a particular food or group of foods.
2. *Hazard characterization*: The qualitative and/or quantitative evaluation of the nature of the adverse health effects associated with biological, chemical, and physical agents that may be present in food. For chemical agents, a dose-response assessment is performed. For biological or physical agents, a dose-response assessment should be performed if the data is obtainable.
3. *Exposure assessment*: The qualitative and/or quantitative evaluation of the likely intake of biological, chemical, and physical agents via food, as well as exposures from other sources if relevant.
4. *Risk characterization*: The qualitative and/or quantitative estimation, including attendant uncertainties, of the probability of occurrence and the severity of known or potential adverse health effects in a given population based on the hazard identification, hazard characterization, and exposure assessment.

3.6.2 RISK MANAGEMENT

The first phase of the RMF consists of "preliminary risk management activities." After a food safety issue has been identified, available scientific information is aggregated into a risk profile that will guide further action. Risk managers may seek additional and more detailed scientific information on an assessment of risks from methodologies such as risk assessment, risk ranking, or epidemiology-based approaches such as source attribution. Ranking using tools that rely on knowledge of risk factors to rank risks and prioritize regulatory controls may be carried out either within or without risk assessments. Epidemiology includes observational studies of human illness such as case control, analysis of surveillance data, and focused research, and is used to apportion risk and contribute to setting risk-based standards. These approaches are often used in combination. If a risk assessment is needed, it can be commissioned from those responsible for that function, with iterative discussions between risk managers and risk assessors to determine the scope of the risk assessment and to decide on the questions it is to answer. Near the end of this preliminary stage, the results of the risk assessment are delivered back to the risk managers and further discussions are generally held about the results and their interpretation. During this "preliminary" phase, good risk communication is important. Communication with external interested parties is often needed to fully identify the food safety issues, obtain sufficient scientific information for risk profiling, and formulate questions to be answered by the risk assessment. Internal communication between risk managers and risk assessors is vital for many reasons, such as to ensure that the scope of the risk assessment is reasonable and achievable, and that the results are presented in a readily understandable form. The second phase of the RMF consists of identifying and evaluating a variety of possible options for managing (e.g., controlling, preventing, reducing, eliminating, or in some other manner mitigating) the risk. As before, effective communication is a prerequisite for success, as information from and the opinions of affected stakeholders—particularly the industry and consumers—are valuable inputs to the decision-making process. Weighing the results of the risk assessment, as well as any economic, legal, ethical, environmental, social, and political factors associated with the risk-mitigating measures that might be implemented, can be a complex task. Economic evaluation of possible risk management interventions enables risk managers to examine the health impacts and feasibility of a proposed intervention relative to its cost. An open and participatory process helps ensure that the final decision is understood and widely supported by those affected by it. When preferred risk management options have been selected, they must be implemented by the relevant stakeholders. In many countries today, industry has the primary responsibility for implementing regulatory standards. However, some nonregulatory risk management options may be selected, such as quality assurance schemes at the farm level, or consumer education packages for food handling in the home. Generally, national food safety authorities must validate and verify implementation of regulatory standards. Once control measures have been implemented, monitoring and review activities should be carried out. The goal is to determine whether the measures that were selected and implemented are in fact achieving the risk management goals they were meant to achieve, and whether they are having any other unintended effects. Both industry and government bodies are likely to be involved in monitoring and reviewing activities. Both sectors usually monitor levels of hazard control, while governments generally carry out health surveillance of the population to determine the level of foodborne illness. If monitoring information indicates a need to review the decision about risk management options, the risk management process can begin a new cycle, with all interested parties participating as appropriate. When dealing with a given specific food safety issue, the RMF can be entered into at any phase and the cyclical process can be repeated as many times as is necessary. What is most important is that appropriate attention is paid to all the phases in the process. More than anything else, application of the RMF represents a systematic way of thinking about all food safety issues that require risk management. The level of intensity of each phase will be matched to the needs presented by each food safety issue and may range from simple, qualitative processes to complex scientific and social evaluations (FAO, 2009).

A summary of the phases and steps in risk management is as follows:

- Phase 1: Preliminary risk management activities
  - Identify food safety issue
  - Develop risk profile
  - Establish goals of risk management
  - Decide on the need for risk assessment
  - Establish risk assessment policy
  - Commission risk assessment, if necessary
  - Consider results of risk assessment
  - Rank risks, if necessary

- Phase 2: Identification and selection of risk management options
  - Identify possible options
  - Evaluate options
  - Select preferred option(s)
- Phase 3: Implementation of risk management decision
  - Validate control(s) where necessary
  - Implement selected control(s)
  - Verify implementation
- Phase 4: Monitoring and review
  - Monitor outcome of control(s)
  - Review control(s) where indicated

3.6.3   RISK COMMUNICATION

According to World Health Organization, risk communication is defined as "an interactive exchange of information and opinions throughout the risk analysis process concerning risk, risk-related factors and risk perceptions among risk assessors, risk managers, consumers, industry, the academic community and other interested parties, including the explanation of risk assessment findings and the basis of risk management decisions." Risk communication is a powerful yet often neglected element of risk analysis. In a food safety emergency situation, effective communication between scientific experts and risk managers, as well as between these groups, other interested parties, and the general public, is absolutely critical for helping people understand the risks and make informed choices. When the food safety issue is less urgent, strong, interactive communication among the participants in a risk analysis almost always improves the quality of the ultimate risk management decisions, particularly by eliciting scientific data, opinions, and perspectives from a cross section of affected stakeholders. Multi-stakeholder communication throughout the process also promotes better understanding of risks and greater consensus on risk management approaches. So given its value, why is risk communication frequently underutilized? Sometimes food safety officials are simply too overwhelmed with collecting information and trying to make decisions to engage in effective risk communication. Risk communication also can be difficult to do well. It requires specialized skills and training, to which not all food safety officials have access. It also requires extensive planning, strategic thinking, and dedication of resources to carry out. Since risk communication is the newest of the three components of risk analysis to have been conceptualized as a distinct discipline, it is often the least familiar element for risk analysis practitioners. Nevertheless, the great value that communication adds to any risk analysis justifies expanded efforts to ensure that it is an effective part of the process. Risk communication is fundamentally a two-way process. It involves sharing information, whether between risk managers and risk assessors, or between members of the risk analysis team and external stakeholders. Risk managers sometimes see risk communication as an "outgoing" process, providing the public with clear and timely information about a food safety risk and the measures in place to manage it (and, indeed, that is one of its critical functions), but "incoming" communication is equally important. Through risk communication, decision makers can obtain vital information, data, and opinions and solicit feedback from affected stakeholders. Such inputs can make important contributions to the basis for decisions, and by obtaining them risk managers greatly increase the likelihood that risk assessments and risk management decisions will effectively and adequately address stakeholder concerns. Everyone involved in a risk analysis is a "risk communicator" at some point in the process. Risk assessors, risk managers, and external participants all need risk communication skills and awareness. In this context, some food safety authorities have communication specialists on their staff. When such a resource is available, integrating the communication function into all phases of a risk analysis at the earliest opportunity is beneficial. For example, when a risk communication specialist can be assigned to the risk assessment team, their presence heightens sensitivity to communication issues and can greatly facilitate communication about the risk assessment that occurs later in the process.

All these activities on risk analysis take place at the organizational or national level. Much of the conceptualization of risk management systems has been taking place at the international level through intergovernmental organizations such as the CAC and the World Organization for Animal Health (OIE). This is not surprising since these organizations are involved in setting the standards for food and animals respectively in international trade. With the diversity of approaches to food production and manufacturing worldwide, it was critical for the CAC to establish general principles and concepts related to food safety risk management decision making that could lead to harmonized international standards for the wide range of foods and potential hazards. The international sharing of ideas has been highly beneficial in identifying the critical role that risk assessment techniques are likely to play in the future to help governments and industry translate public health goals into meaningful risk management programs. Some of the critical CAC documents related to the adoption of risk management approaches are as follows:

- "Working Principles for Risk Analysis for Food Safety for Application by Governments" (CAC, 2007b)
- "Working Principles for Risk Analysis Application in the Framework of the Codex Alimentarius" (CAC, 2007c)
- "Principles and Guidelines for the Conduct of Microbiological Risk Assessments" (CAC, 1999)
- "Principles and Guidelines for the Conduct of Microbiological Risk Management (MRM)" (CAC, 2007a)

These documents deal in details about the risk analysis and are essential for anyone involved in developing food safety risk management systems.

3.7 RISK ANALYSIS AND HAZARD ANALYSIS AND CRITICAL CONTROL POINTS

No discussion of food safety risk management would be complete without considering Hazard Analysis and Critical Control Points (HACCP), the risk management system most widely used with food. Despite the fact that HACCP has been almost universally adopted, there has been surprisingly little discussion about many of the concepts underpinning HACCP and how it fits into the RMF. In many ways, HACCP has suffered for being ahead of its time. HACCP articulated many of the principles associated with systems thinking long before systems thinking was formalized as a problem-solving approach. HACCP is clearly a "semi-quantitative" risk management system that is based on a largely qualitative assessment of hazards. The difference between an HACCP hazard assessment and a risk assessment has long been discussed and remains somewhat controversial. It is clear that selection of "significant hazards" involves consideration of the likelihood and severity of a hazard, implying a risk evaluation. However, the "risk decisions" reached in developing a HACCP plan are often nontransparent and not fully supported by an adequate assessment. HACCP has largely remained unchanged in its approach over the last 40 years. This may, in part, reflect its adoption for regulatory purposes. While HACCP has clearly been a beneficial regulatory tool for bringing a uniformity of approach to food safety, it has also had the unintended consequence of discouraging the evolution of HACCP as a risk management system (CAC, 1995). This has been discussed in detail in Chapter 2, as it has become a basic tool for achieving food safety. HACCP is a risk-based, preventive food safety system used by the food industry. It is a primary risk management system that provides a systematic preventive approach to food safety from various types of hazards in production processes that can cause the finished product to be unsafe. Systems are designed accordingly to reduce these risks to a safer level. Every country has given due importance to HACCP in its food safety control programs and regulations. The United States has given due importance to HACCP in its newly notified rules under the Food Safety Modernization Act (FSMA). Creating a culture of food safety in any organization places emphasis on the Hazard Analysis and Risk-Based Preventive Control (HARPC) system. These rules apply to certain unintentional hazards such as microbiological, chemical, physical, or radiological hazards. It is an extension of HACCP as it additionally includes radiological hazards. Hazard research is another essential component of HARPC. The industry should be able to demonstrate to regulators and auditors the hazardous potential of any such hazard. Efforts are being made to improve upon the existing HACCP approach so as to create an efficient risk-based preventive food safety program. However, the basic principles of HACCP remain the same even when certain operational changes and additions are made at every level. The basic structure and criticalities of HACCP consist of

1. *Identification of hazards*: There is a need to have continuous evaluation and monitoring to identify hazards, as these hazards can change over periods of time. There are possibilities for new additions and deletions in criticality. Every operation and every activity is unique and so are the hazards, and expectation of uniformity may not be appropriate.
2. *Preventive controls innovations*: Prevention requires the control and consistency of raw materials, the processing environment, and process steps. Controls designed and implemented may not be suitable for all situations and all times. Innovations in control systems and designs are essential.

As mentioned above, every operation, hazard, and risk is unique, and specific limits and activities to control these are essential in creating preventive controls. Innovations in control measures are the answer.

3. *Reliability and validity*: Reliability is another important part of the food safety program. It is achieved by ensuring compliance with the preventive approach plan that the organization has established. Validation is also equally essential. Scientific literature, writings published in scientific journals, university studies, and similar studies can be used for validation.

With the emergence of risk management metrics, there is substantial interest on the part of a number of scientists in identifying how FSOs and performance objectives (POs) can be used to help convert HACCP into a risk-based food safety system. It is clear that being able to more directly tie the stringency of an HACCP system to public health outcomes would be highly beneficial for harmonizing both national and international standards, and would lead to more transparent food safety decision making. In particular, it would help address some of the long-standing limitations associated with HACCP. For example, the establishment of critical limits is an area that has eluded sound guidance, in part because there have been insufficient means for relating the level of stringency required at individual critical control points to expected food safety outcomes.

The inclusion of risk management and risk assessment concepts, techniques, and tools in HACCP may also help dispel some of the misconceptions about HACCP. For example, there is a general impression that HACCP cannot be applied to food systems where there is not a clear intervention step that eliminates the hazard. Likewise, there have been questions about whether HACCP can be used effectively in the farm and retail segments of the food chain, or whether it can be effectively applied to food defense issues. However, as a systems approach to managing food safety risks, HACCP should be viewed relative to its general framework—that is, managing food safety risks by gathering knowledge about the hazards associated with a food, determining the risk that these hazards represent, developing risk mitigations that control the hazards to the degree required, and developing metrics that assess whether the required level of control is being achieved. With new decision tools that are being developed and applied to food safety issues, it should be feasible to greatly enhance HACCP systems. With the power of software applications, it should be possible to make these tools available in user-friendly formats suitable for all segments of the food industry.

3.8 STRATEGIES AND CONTROLS FOR FOOD SAFETY

We have already discussed how, in the history of food safety, various strategies and controls have been used from time to time to achieve food safety. The methods adopted were based on experience and differed from place to place. Over the years, situations and needs have changed, and now consumers are spread out and demanding safe and high-quality food. Science, which has become part of life, has also entered into the field of food safety. Strategies have changed and so have the control systems. Different strategies, which are used in combination or in isolation, have

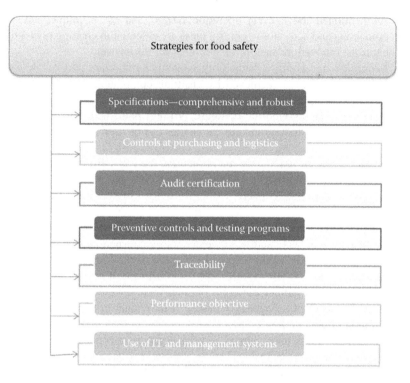

FIGURE 3.5   Strategies for food safety.

been always present in one form or another. A few of these were developed in the past and are growing in importance. These seven strategies, which have been illustrated in Figure 3.5, are

1. *Comprehensive and robust specifications*: The setting of standards is a basic step before any control system can be conceptualized. Standards should be based on scientific data and analysis. The availability of complete health surveillance data, disease outbreak information, epidemic data, daily dietary intake information, chemical safe limit studies, and shelf life studies is a must before standard-setting exercises can be initiated. Specifications are developed after conducting comprehensive research into food safety, product development, and risk management. Hazards are identified and addressed. Standards and specifications must also keep continuity and consistency. Without well-thought-out specifications, planning and designing for a quality product cannot be done. Standards also lay down the foundation for the further action plan in any organization. Specifications are precursors to comprehensive purchasing agreements and supplier review programs. These also help in deciding the production facility requirements and marketing strategies.

2. *Purchasing and logistics controls*: This starts with the collection of raw materials. If raw materials are not of good quality, products cannot be safe and of the desired quality, so it becomes essential to have raw materials that are safe and conform to the specifications. This is one area that must not be overlooked by the food safety professional. Input from the food safety professional is vital regarding what is required of the supplier to minimize risk. The foremost prerequisite is that the products comply with all of the rules and regulations regarding food or ingredients. It is better to have a system for auditing by both a certified third party and the purchasing company. It must be ensured that good agriculture and manufacturing practices were followed, and that the sampling protocol is representative of the lot size. Care must be taken that the sample is tested from the actual lot that is being purchased. The credibility of the laboratory or agency that is conducting the tests must be established. Similarly, the conditions of transport and storage are to be monitored. It has happened many times that the material purchased was of good quality but there were problems during transportation or storage that rendered the food unsafe, so appropriate controls during purchasing and transportation are essential.

3. *Audits and certifications*: Use of certified auditing programs is one of the control methods used frequently these days. Audits are done internally on a regular basis for high-risk products. Corrective actions are also indicated from time to time as per the requirements. Third-party audits by certified agencies are also gaining importance. The frequency of an audit needs to be assessed. Audit frequency is often dictated by the "risk" of the product and the historical data of working with the supplier. There are complexities in the food trade, especially with incidents of adulteration for economic fraud. There are cultural differences in dealing with such frauds. Historical information or recent intelligence regarding suppliers is also very useful. Sometimes internal audits are also conducted as these help to establish and improve relationships. The challenge is covering the cost of the audit and the resources with which to conduct it.

4. *Preventive controls and testing programs*: There are various preventive controls systems that are used to ensure that contamination can be avoided and food is kept safe. They start with the initiation of agricultural practices and continue even after the product leaves the production facility. Now even packaging materials have been designed such that these controls can be monitored after the product is marketed and until it is consumed. There are good agricultural practices, good manufacturing practices, and good hygienic practices. Use of intelligent packaging is also gaining importance. The most widely used control system during the production process is HACCP, and using a HACCP-based approach reassesses food safety systems in consideration of intentional and unintentional risks. A protocol is followed to properly identify pertinent hazards and establish preventive controls that are commensurate with current programs and production plans. Companies normally sample and test incoming products, but the control does not end there as testing is not a substitute for a well-developed HACCP program.

5. *Traceability*: It is important to develop and implement a traceability program. Today, all establishments amenable to the Bioterrorism Act, 2002, must be able to track and trace products one step forward and one step backward. A recent study conducted by the Office of the Inspector General in 2008 on "Traceability in the Food Supply Chain" revealed that only 5 of 40 assorted food products purchased from retail establishments were traceable throughout the entire supply chain. 31 of 40 were traceable to locations that "likely" handled the product, but not all locations could provide the required lot-specific records. The report went on to state that 59% of the food facilities they investigated did not meet the FDA's records requirements for transporters, suppliers, and customers.

These strategies and control systems are used for minimizing risk within the global food supply chain. Reliability and trust play a major role, both for manufacturers and consumers. This starts with a clear understanding of the vulnerabilities both upstream and downstream across supply chains, and flows to preventive controls and programs that are verified and validated with adequate frequency. Programs designed to respond to a crisis in a timely and efficient way to protect the product, public health, and the food industry are key to the planning and preparedness process.

6. *Performance objectives*: One of the risk management tools that has helped to facilitate the move to quantitative risk management has been the emerging concepts related to food safety objectives (FSOs) – performance objectives (POs) that were introduced by the International Commission on Microbiological Specifications for Foods (2002) and then elaborated by a series of FAO/WHO expert consultations and the CAC Committee on Food Hygiene. These tools and concepts are dramatically changing the way that more traditional risk management metrics are being developed. The key to this approach is to use risk-modeling techniques to relate levels of exposure to the extent of public health consequences that are likely to occur. This information is then used to determine the levels of hazard control that need to be achieved at specific points along the food chain. This approach has the distinct advantage of including a clearly articulated level of control based on risk and flexibility in terms of the strategies and technologies that can be used to control foodborne hazards. It also provides a means of comparing the effectiveness and levels of control that can be achieved by focusing prevention/intervention efforts at different points along the food chain. For example, this approach is beginning to evaluate questions such as whether fresh-cut produce manufacturers would be better served by emphasizing on-farm prevention or postharvest interventions. The use of risk management metrics concepts is also having a significant impact on the way that microbiological criteria are being developed. Two key examples are the CAC microbiological criteria developed recently for powdered infant formula and *Listeria monocytogenes* in ready-to-eat foods.

7. *Use of IT and management systems*: Technology enables companies to adopt an integrated approach to food safety programs. The solutions available in the market are packed with strong capabilities and functionalities

that empower companies to adopt a workflow-based approach, automate and streamline key processes, eliminate silos, and seamlessly achieve the desired results from the risk-based preventive food safety program. These include

a. *Compliance management*: Companies can leverage technology to centrally manage compliance standards and requirements mapped to relevant countries, business units and processes, products, and suppliers. Central repositories with strong document-management functionalities help with recordkeeping.

b. *Supplier and product information management*: Sophisticated web-based solutions simplify the process of capturing and managing supplier and product information, including subsuppliers, lab testing, verification and validation on preventive control, country of origin, and transit information.

c. *Hazard analysis and risk management*: Technology helps automate the end-to-end hazard analysis and risk management processes. Solutions include templates to facilitate hazard classification and tools such as risk heat maps, risk calculators, and scorecards for hazard analysis and risk assessment. These help in enhancing the process of identifying, evaluating, and prioritizing food quality and safety risks, including supplier and supply chain risks. The solutions are equipped with risk rating and ranking capabilities, and help to determine the impact, likelihood, severity, and frequency of hazards and risk in a simple and consistent manner. Companies can also confidently manage a centralized hazard database for easy reference, and mitigate these risks and hazards.

d. *Preventive controls management*: Advanced solutions help in creating and managing preventive controls for CCPs and non-CCPs, including process controls, food allergen controls, sanitation controls, recall plans, environment monitoring, supplier verification, and training. They can also link controls to identified hazards and centrally manage scientific and technical information to determine if controls are adequate. They can leverage industry-based methodologies for validation of preventive controls such as product testing and environment testing.

e. *Audit and inspection management*: These solutions help in conducting audits, including internal, quality, and safety audits. Companies can also efficiently conduct controls monitoring. Cutting-edge audit management solutions provide companies with the ability to centrally manage audit resources, schedules, and checklists. Offline audit solutions pave the way for the gathering of audit data even in remote locations that are not connected to the central corporate network and have no Internet access.

A risk-based preventive approach to food safety is critical. An integrated technology solution can help maintain a central repository for managing multiple compliance standards and requirements, thus ensuring compliance with the law. At a time when the food industry is under pressure to regain consumer trust following a series of food safety incidents worldwide, a preventive approach to food safety can go a long way in ensuring safe food, regaining consumer trust, and protecting the brand images of companies operating in this field. With a preventive approach in place, organizations can ensure quality products and minimal numbers of product failures and recalls, and minimize spending on such issues.

REFERENCES

Ackerley N., Sertkaya A., and Lange R. (2010). Food transportation safety: Characterizing risks and controls by use of expert opinion. *Food Protection Trends* 30(4): 212–222.

Batz M., Hoffmann S., and Morris J.J. (2011). Ranking the risks, the 10 pathogen-food combinations with the greatest burden on public health, University of Florida, Emerging Pathogens Institute, Technical report. https://folio.iupui.edu/bitstream/handle/10244/1022/72267report.pdf, Accessed on October 10, 2015.

Buchanan R.L. (2010). Bridging consumers' right to know and food safety regulations based on risk assessment. In *Risk Assessment of Foods*, Lee C.H., ed., Korea: KAST Press, pp. 225–234. www.cdc.gov/pulsenet/.

Buzby J.C., Unnevehr L.J., and Roberts D. (September 2008). *Food Safety and Imports: An Analysis of FDA Food-Related Import Refusal Reports, EIB-39*, U.S. Department of Agriculture, Economic Research Service.

CAC (Codex Alimentarius Commission). (1995). Guidelines for the application of the Hazard Analysis Critical Control Point (HACCP) system (CAC/GL 18-1993). In *Codex Alimentarius, Vol. 1B, General Requirements (Food Hygiene)*, Whitehead, A.J., and Orriss, G., eds., Rome, Italy: FAO/WHO, pp. 21–30.

CAC (Codex Alimentarius Commission). (1999). Principles and guidelines for the conduct of microbiological risk assessments. (CAC/GL 30-1999). www.fao.org/docs/eims/upload/215254/CAC_GL30.pdf, Accessed on October 10, 2015.

CAC (Codex Alimentarius Commission). (2007a). Principles and guidelines for the conduct of Microbiological Risk Management (MRM). (CAC/GL 63-2007). http://std.gdciq.gov.cn/gssw/JiShuFaGui/CAC/CXG_063e.pdf, Accessed on October 10, 2015.

CAC (Codex Alimentarius Commission). (2007b). Working principles for risk analysis for food safety for application by governments. (CAC/GL 62-2007). www.fao.org/input/download/standards/10751/CXG_062e.pdf, Accessed on October 10, 2015.

CAC (Codex Alimentarius Commission). (2007c). Working principles for risk analysis application in the framework of the Codex Alimentarius. (CAC/GL 62-2007). http://www.fao.org/docrep/006/y4800e/y4800e0o.htm, Accessed on October 10, 2015.

Chen C., Zhang J., and Delaurentis T. (2014). Quality control in food supply chain management: An analytical model and case study of the adulterated milk incident in China. *International Journal of Production Economics*, 152(C): 188–199.

Convertino M. and Liang S. (2014). Probabilistic supply chain risk model for food safety. GRF Davos Planet @ Risk, Vol. 2 Number 3, April, 2014. Features Food safety—AIB updates, September–October 2014.

FAO. (2009). Food safety risk analysis: A guide for national food safety authorities, FAO, Rome, Italy, Food and Nutrition Papers 87.

FAO/WHO. (2006). *Food Safety Risk Analysis. A Guide for National Food Safety Authorities.* Rome, Italy: FAO/WHO. ftp://ftp.fao.org/docrep/fao/009/a0822e/a0822e00.pdf.

FDA. (2011). Pathway to global safety and quality. http://axendia.com/FDA-Pathway-to-Global-Product-Report.pdf, Accessed on October 12, 2015.

Gary A., Henry C.W., and Feldstein F. (December 2011). The food safety challenge of the global food supply chain. Food Safety Magazine, www.foodsafetymagazine.com/magazine-archive1/december-2011january-2012/the-food-safety-challenge-of-the-global-food-supply-chain/.

International Commission on Microbiological Specifications for Foods. (2002). *Microorganisms in Foods 7: Microbiological Testing in Food Safety Management.* New York: Kluwer Academic/Plenum Publishers.

Nepusz T., Petroczi A., and Naughton D.P. (2009). Network analytical tool for monitoring global food safety highlights china. *PLoS ONE* 4: e6680.

Sinha S. and Lopez S. (2014). Adopting a risk-based preventive approach to food safety across the complex supply chain. www.aibonline.org/aibOnline_/www.aibonline.org/newsletter/Magazine/Sep_Oct2014/RiskBasedApproach.pdf, Accessed on October 12, 2015.

WHO. (2010). *World Health Statistics 2010.* WHO Library Cataloguing-in-Publication Data. www.who.int/whosis/whostat/EN_WHS10_Full.pdf, Accessed on October 12, 2015.

4 Food Quality

4.1 INTRODUCTION

In recent times, food quality has been gaining importance all over the world, even when the reasons for this may differ from country to country. Every country, or sometimes even the regions within a country, has certain specific issues that go beyond food quantity and safety. Focus on food security has been central in most countries of Asia and Africa, but this is shifting in emphasis from quantity to quality, as malnutrition and undernutrition are issues that are not only confined to quantity but are impacted by the quality of the food as well. In countries from the American and European continents, quantity may not be an issue of concern, but safety and nutrition have been gaining more importance. This is because poor diets are resulting in various lifestyle diseases and obesity. Food may be safe, but it still may not be able to support a healthy body. Food safety is but one aspect of the overall quality of food, an issue that is gaining increasing attention all over the world. Though availability and access to food remain critical components of food security in most parts of the world, especially in Asia and Africa, the focus is shifting from quantity to quality. High-profile cases involving food safety have drawn the most attention, but obesity and undernutrition/malnutrition are of equal concern. Food quality has become a rising concern across the globe, not least because food safety scandals have been a regular occurrence in recent years. There is a growing recognition that the focus on quantity has been at the expense of quality. In the 1970s and 1980s, most parts of the world, especially the densely populated Asia, were addressing the issue of whether there was enough food to feed the increasing population. Emphasis was placed on controlling the population and also improving production and productivity in agriculture. This race for increased production resulted in the increased and indiscriminate use of pesticides and antibiotics. This so-called green revolution also resulted in changing food patterns in most of the countries, and a variety of foods and local crops were replaced by uniform crops, creating more nutritional imbalances. One challenge of this focus on high productivity and yield is that the nutrition side has been neglected. Undernutrition and malnutrition are usually terms used in reference to those who do not get enough to eat, or do not get enough of the necessary nutrients. But imbalances in food habits and patterns are the reasons for many lifestyle diseases as well. Where food is adequate in quantity, the problem of balanced diets is causing concern. Obesity is becoming a major issue in many countries, one that is increasingly linked to the consumption of more processed foods that are high in salt, fat, and sugar. More affluent nations such as the United States, Australia, New Zealand, South Korea, and Japan are starting to think about obesity in terms of preventative health-care. Some of the countries that are struggling with issues of food security such as India and the Philippines are also experiencing an obesity epidemic. High levels of income inequality have driven the situation, whereby the excess intake of calories coexists with undernutrition. A large number of deficiency diseases are also causing concern, as these result in serious medical conditions. Fortification of food to address iron, iodine, or any other kind of deficiencies is becoming very common to remove imbalances in the diet. Despite healthy economic growth rates and an overall decline in poverty in South Asia and Africa, micronutrient malnutrition remains pervasive—and not just among the poorest or most vulnerable (United Nations, 2011). Food quality has become a rising concern for people who wish to have a healthy and balanced diet, and also where food safety scandals have been a regular occurrence in recent years. Therefore, "food quality" is a comprehensive term that includes all the required parameters essential for maintaining the safety and security of food.

The *Economist* (2013) studied the aspects of food quality in Asia and reported many cases that had an impact on trade but ultimately resulted in improving the quality of food products. Five years on from an incident in which six babies died and over 300,000 became ill from melamine-tainted infant formula, parents in mainland China remain distrustful of local milk powder brands. Their fear has created supply issues in many markets around the world, notably Hong Kong, where the government felt it necessary to limit outbound travelers to a quota of just two cans each. In March and April of 2013, more people were detained in Hong Kong for smuggling milk powder—all of it destined for mainland China—than were arrested for drug trafficking in the whole of 2012. It is not just milk powder that is cause for concern. In 2013, thousands of dead pigs were found floating in Shanghai's Huangpu River, and 900 people were arrested in a major crackdown on "meat-related crimes," including selling rat meat in the guise of mutton. In India, reports of traders ripening fruits with "masala spice"—or calcium carbide, a known carcinogen—reflect the underdevelopment of infrastructure for food storage and transport. A 2011 criminal investigation into the deaths of 143 people from bootleg liquor laced with methanol in West Bengal is a stark reminder that around 90% of food production in the country remains informal and based in poor urban areas. At the other end of the spectrum, detection of a postharvest fungus halted exports of New Zealand apples to China in September 2013, and the spotlight fell on the country's world-class dairy industry in August 2013 when its main exporter, Fonterra, reported the discovery in its supply chain of a strain of bacteria that can cause botulism (TELARC). Following a global recall and a blanket ban on its products in China, further investigations proved the case to be a false alarm, yet the scare indicated that even the best systems are not immune to breaches and that effective crisis communications and risk management strategies are important aspects of managing food safety across increasingly complex global supply networks (Van Nederpelt, 2009).

Food quality is the quality characteristics of food that are acceptable to consumers. Acceptability has wide connotations. Normally we confine it to sensory evaluations, but there are also concerns about safety and security as well. These include the expectations of the customer, the declarations of the producers, nutritional requirements, legal requirements regarding safety, and other issues. To sum up, food quality includes the following:

1. *Customer-oriented sensory-evaluated attributes*: These are evaluated by the sensory faculties. The attributes are as follows:
   a. External factors: The appearance of the food, including size, shape, color, gloss, and consistency. In fact, in most cases the consumer decides to buy a product based only on appearance.

b. Texture and flavor: The look, feel, appearance, and flavor of the product attract customers as they expect a certain quality based on their previous experience.

c. Internal factors: Chemical, physical, and microbial factors that are known to the consumer only after they take the food. In this case the consumer also has to rely on their previous experience. Actual properties will be known through detailed examination by analysts in laboratories.

2. *Food safety and regulatory concerns*: Food quality is an important requirement in the food regulations of all countries because food consumers are susceptible to any form of contamination or adulteration that may occur during any stage. Many consumers also rely on labels and declarations to find specific information that is relevant to them. Some are interested to know what ingredients are present, sometimes due to dietary and nutritional requirements (e.g., kosher, halal, or vegetarian) or medical conditions (e.g., diabetes or allergies). Besides ingredient quality, there are also sanitation requirements. It is important to ensure that the food processing environment is as clean as possible in order to produce the safest possible food for the consumer. Minimum standards in food quality and food safety are regulated through regulations and laws. Under these regulations/laws, the enforcement agencies of the government inspect food products and facilities and take samples to check the facility or product for compliance with the standards. Also, members of the public complain to regulatory bodies, who take complaint samples and samples used to routinely monitor the food marketplace to public analysts. Public analysts carry out scientific analysis on the samples to determine whether the quality is of a sufficient standard. Food quality also deals with product traceability (e.g., of ingredient and packaging suppliers), should a recall of the product be required. It also deals with labeling issues to ensure the correct ingredients and nutritional information are listed.

3. *Voluntary quality standards*: There are a number of voluntary standards for the quality of food products that have been set by associations, private bodies, and also by governments. Governments have taken the initiative to create such voluntary standards because compulsory standards set up through regulations take care of safety requirements and also minimum quality requirements, but if the consumer demands better quality then they can go for such certified food products.

4.2 FOOD QUALITY EVALUATION

Evaluation of quality is essential in order to confirm if the product finally produced meets the standards fixed for it. There are certain tests that are performed to evaluate food quality. These tests are performed as per the requirements of the type of evaluation. Samples of the product are taken and prepared accordingly. There are fixed protocols and procedures for taking and preparing samples, depending upon the requirements. Food products can be analyzed to see how they conform to regulatory standards or sometimes to voluntary standards or parameters fixed by the

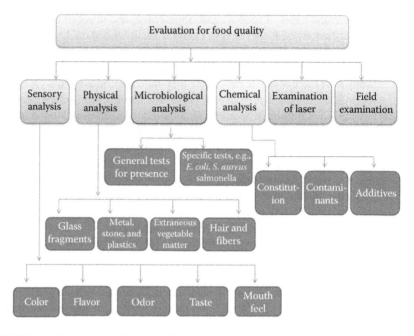

FIGURE 4.1 Evaluation of food quality.

company itself. The following types of tests/evaluations, as illustrated in Figure 4.1, can be done to verify and confirm the quality of food:

1. Sensory analysis
2. Physical analysis
3. Microbiological analysis
4. Chemical analysis
5. Examination of label
6. Field examination

4.2.1 SENSORY ANALYSIS

Sensory analysis (or sensory evaluation) is a scientific discipline that applies the principles of experimental design and statistical analysis to the use of human senses (i.e., sight, smell, taste, touch, and hearing) for the purposes of evaluating consumer products. This discipline requires panels of human assessors on whom the products are tested, and their responses are recorded. By applying statistical techniques to the results, it is possible to make inferences and insights about the tested products. There are several types of sensory test. The most classic is the sensory profile. In this test, each taster describes each product by means of a questionnaire. The questionnaire includes a list of descriptors (e.g., bitterness, acidity, etc.). The taster rates each descriptor for each product depending on the intensity he perceives in the product (e.g., 0 = very weak to 10 = very strong).

The main sensory characteristics of food that can be evaluated are

1. *Appearance*: The surface characteristics of food products contribute to their appearance. Interior appearance can also be evaluated. Perception of size, shape, color, and transparency can be judged by external appearance. Lumps in a pudding or gravy, which are not desirable, can be judged by the eye.
2. *Color*: In addition to giving pleasure, the color of foods is associated with other attributes. Color is used as an index for the quality of a number of foods. The ripeness of fruits like bananas, mangos, tomatoes, and guavas can be assessed by their color.
3. *Flavor*: The flavor of food has three components: odor, taste, and a composite of sensations known as "mouth feel."
   a. *Odor*: The odor of food contributes immeasurably to the pleasure of eating. Aromas are able to penetrate even beyond the visual range when comparatively volatile compounds are abundant in the food.
   b. *Taste*: We value food for its taste. The sensation known as "sourness" is associated with hydrogen ions supplied by acids like vinegar. A salty taste is due to ions of salt. Sweet tastes are related to the presence of organic compounds.
   c. *Mouth feel*: Texture and consistency can be found out by mouth feel, as well as heat or a burning sensation of pepper. Mouth feel depends upon the temperature, texture, and tenderness of food products.

Sensory evaluation can be used to

- Compare similarities/differences in a range of dishes/products
- Evaluate a range of existing dishes/products
- Analyze food samples for improvements
- Gauge responses to a dish/product (e.g., acceptable versus unacceptable)
- Explore specific characteristics of an ingredient or dish/product
- Check whether a final dish/product meets its original specification
- Provide objective and subjective feedback data to enable informed decisions to be made

4.2.2 Physical Analysis

Customers will not accept finding a "foreign body" in their food. Sometimes the object is a perfectly natural component of the food, but in all cases it is important to find out what it is and how and when it got there. Physical analysis of the food is conducted for the identification of an adventitious and deliberate contaminant, helping to pinpoint its source and advising on preventative solutions. Some of those that can be spotted are

1. *Glass fragments*: Light microscope examination for features such as size, color, curvature, and deposits is followed by X-ray microanalysis in the scanning electron microscope. This quick, nondestructive test gives a spectrum of the elements found in a sample, allowing for rapid identification.

2. *Metal, stone, and plastics*: These and other nonbiological materials are subjected to a similar examination process to that used for glass samples. X-ray microanalysis can distinguish between different metals, including the different types of stainless steel. With plastics, other techniques such as Fourier transform infrared spectroscopy (FT-IR) can be used to "fingerprint" the material. This technique is also invaluable for many pharmaceutical tablets and capsules.

3. *Extraneous vegetable matter*: This is often identified by light microscopy to demonstrate its cellular structure, using various histochemical stains to identify the chemical nature of the material. Similar approaches can be used to identify other similar samples such as rodent droppings.

4. *Hair and fibers*: These are usually identified using light microscopy and compared with a reference collection of known samples.

4.2.3   MICROBIOLOGICAL ANALYSIS

In products in which microorganisms can survive and grow, routine microbiological analysis is important to confirm that manufacturing control mechanisms are effective. It is also necessary for the checking of raw material quality (e.g., to confirm that it is within specification), or for investigating customer complaints. Microbiological test procedures for the examination of foods and beverages have been standardized and regulated, but nearly every country has its own regulations. These are conducted using various culture media and base ingredients (e.g., chromogenic media for the differentiation of microbes based on colony color), and tests are conducted through biochemical, immunological, and molecular biological methods for pathogen detection, identification, and confirmation. Microbiological testing on food products includes the presence/absence of pathogens, and total coliform and aerobic plate counts. The total coliform count determines the number of coliform bacteria. These bacteria are commonly found in the intestines of warm-blooded animals or in the environment—for example, soil, water, and grain. High coliform levels can serve to indicate unsatisfactory processing and sanitation and the possibility of other pathogens proliferating since the conditions for growth are similar. The aerobic plate count determines the total number of aerobic bacteria in a food, and is useful for indicating the microbiological quality of a food product, the potential spoilage in perishable products, and the sanitary conditions under which the food was processed. There are a few pathogens for which specific tests must be conducted to check for their presence. *Staphylococcus aureus* counts determine the presence of this organism in food, which can indicate poor food handling practices by workers and inadequate sanitation and temperature control. The presence of the *Salmonella* spp. organism in an ingested food can result in food poisoning. Symptoms include abdominal pains, diarrhea, nausea, and vomiting 2–5 days after consumption. Other tests that can be performed determine the presence/absence of and/or identify yeast/molds, *Bacillus cereus*, foreign substances in aerated products, and fungi.

4.2.4 CHEMICAL ANALYSIS

Understanding the chemical makeup of products is essential for a variety of reasons—for providing accurate health claims, nutrition labeling, and allergen warnings, to name but three—but there are many other aspects of food chemistry that need to be appreciated so that problems can be addressed when they arise, and even prevented from occurring. As well as the "normal" nutritional composition of food, we need to know about the levels of natural toxicants and allergens, and of potential contaminants such as pesticides and packaging migrants.

The nutritional composition of food is of major significance to the consumer—and to the authorities—and so it is important to know what nutrients are in our products. These include

- Fats, including saturated, monounsaturated, polyunsaturated, trans fats, and omega 3 and 6
- Proteins
- Carbohydrates, including sugars, starch, inulin, polysaccharides, and polyols
- Mineral matter and ash
- Dietary fiber
- Salt
- Meat and fish content
- Alcohol
- Minerals, trace metals, and other trace elements
- Vitamins
- Antioxidants
- Additives, including preservatives, antioxidants, sweeteners, colors, emulsifiers and stabilizers, and gelling agents
- Organic acids such as lactic, citric, malic, and so on

Many foods naturally contain chemicals that are either toxic to a greater or lesser degree or contain allergenic proteins. The range of potential problems is wide, but analyses are carried out to see if a hazard exists in the products. Among the many tests that can be performed are those for

- Phytohemagglutinins/lectins
- Allergens, including eggs, dairy (e.g., casein, whey, lactose), nuts, gluten, sulfites, histamines, and sesame
- Mycotoxins, including ochratoxin, aflatoxin, zearalenone, patulin, and sterigmatocystin
- Furans
- Acrylamide
- Metals (e.g., lead, cadmium, mercury, arsenic)
- Glucosinolates
- Glycoalkaloids
- Alkaloids (e.g., caffeine and theobromine)

Foods can often become contaminated with chemicals that should not be present. These may arise from the growing environment (e.g., the soil or the air) or from processing or packaging. These include

- Heavy metals and trace elements, including iron, lead, copper, zinc, tin, cadmium, arsenic, and mercury
- Mycotoxins, including ochratoxin, aflatoxin, zearalenone, and patulin
- Polyaromatic hydrocarbons
- Polychlorinated biphenyls
- Monochloropropanediol (3-MCPD) and dichloropropanol (1,3-DCP)
- Furans
- Phthalates
- Benzene
- Nitrates and nitrites
- Solvents
- Acrylamide
- Illegal colors (e.g., Sudan I)
- Packaging contaminants
- Food taints, including wine taints
- Antibiotics
- Pesticide analysis and taint identification
- Melamine

As the ability to detect pesticide residues in minutely small amounts increases, so the legal tolerances for residues get even stricter. Many maximum levels are set at the "limit of quantification," so it is important to use an analytical laboratory that can use the most sensitive techniques. It is important to carry out the appropriate methods of pesticide analysis; as well as looking for individual pesticides, we can undertake multiresidue screening, thus saving both time and money.

Packaging must not transfer its constituents into food to the detriment of the food's quality. In most countries legislation is very strict, especially regarding heavy metals and plastic monomers. To test compliance with legal requirements for both heavy metal transfer and plastic migration from packaging into food, a number of tests are conducted to investigate the migration of the following contaminants:

- Formaldehyde
- Acrylonitrile
- Vinyl acetate
- Bisphenol A
- BADGE, NOGE, BFDGE
- Photoinitiators
- Phthalates

4.2.5 EXAMINATION OF LABELS

Labels are panels found on packages of food that contain a variety of information about the nutritional value of the food item inside. There are many pieces of information that are standard on most food labels, including serving size, number of calories, grams of fat, included nutrients, and a list of ingredients. This information helps people who are trying to restrict their intake of fat, sodium, sugar, or other ingredients, or those individuals who are trying to get enough healthy nutrients such as calcium or vitamin C. The label provides each item with its approximate percent daily value. Apart from this, information on weight, volume, price, and so on is also given on the label. There is certain information that is given by the manufacturer for advertising the product. Other information, which is mostly mandatory, is given to inform consumers about the product.

4.2.6 FIELD EXAMINATION

A field examination is an on-the-spot examination or field test performed on a product to support a specific decision. It may be conducted on products discharged from vessels onto wharves, piers, pier sheds, and other locations; products in trucks, trains, freezers, and containers at border entry points; or on products set aside for examination by the U.S. Food and Drug Administration (FDA). Some compliance program guidance manuals do not address field examinations. Nevertheless, field examinations are appropriate for certain problems and/or commodities and should be conducted.

A field examination involves actual physical examination of the product for such things as

- Quantity, which should correspond to the quantity declared on the shipping documents
- In-transit or storage damage
- Inadequate storage temperature conditions
- Rodent or insect activity
- Lead in ceramicware (through the quick color test and rapid abrasion test)
- Odors uncharacteristic of the product or indicating spoilage
- Nonpermitted food and/or color additives
- General label compliance

When conducting a field examination, compare documents provided by the filer/ importer to what is physically available during your inspection.

4.3 MANAGEMENT AND MAINTENANCE OF QUALITY

Now that we know what quality is, and the requirements or standards for regulatory purposes or otherwise so that we have a way to measure quality, the question is: how do we manage quality? Or, more specifically, how do we manage and maintain quality so that we are sure that the food product produced has the desired characteristics?

Quality management is not a very old phenomenon. "Quality control" and "quality assurance" were the terms we used earlier, maybe with different meanings as conforming to standards was the main aim. Nowadays, quality management is being adopted in each industry and all products. Advancements and innovations have allowed consumers to choose goods that meet higher quality standards than normal goods. In old societies, the production of arts and crafts was the responsibility of master craftsmen or artists; these masters would lead their studios and train and supervise others. Production was limited and the main artist was able to maintain quality as the number of items was limited. The Industrial Revolution brought major changes to production systems, and mass production replaced limited production. Emphasis was placed on producing the same products on a large scale and this culminated in including aspects like standardization and improved practices. Food production on a mass scale came a little later, as mass production was first used in the production of other products. Similarly, quality management was first used in the production of machines and similar products, and this is the reason that innovations in the field of quality management and efficiency improvement came from the auto industry. Henry Ford was important in bringing process and quality management practices into operation on his assembly lines; he standardized processes and products. In Germany, Karl Friedrich Benz, often credited with being the inventor of the motor car, pursued similar assembly and production practices, although real mass production was properly initiated by Volkswagen after the Second World War. From this period onwards, North American companies focused predominantly upon production with lower costs and increased efficiency. Walter A. Shewhart made a major step in the evolution toward quality management by creating a method for quality control for production, using statistical methods. Quality leadership has witnessed a major change over the past five to six decades. After the Second World War, Japan decided to make quality improvement a national imperative as part of rebuilding their economy, and sought the help of well-known experts in quality management like Shewhart, Deming, and Juran. Dr. W. Edwards Deming is known for his management philosophy that establishes quality, productivity, and competitive position. He has formulated 14 points of attention for managers, which are a high-level abstraction of many of his deep insights. They should be interpreted by learning and understanding the deeper insights. These include key concepts such as

- Break down barriers between departments
- Management should learn their responsibilities and take on leadership
- Supervision should help people, machines, and gadgets to do a better job
- Improve constantly and forever the system of production and service
- Institute a vigorous program of education and self-improvement

Due to special efforts to manage quality over time, Japan, which was synonymous with cheapness and low quality, began to implement successful quality initiatives, and Japanese goods achieved very high levels of quality in its products from the 1970s onward (Selden, 1998).

Customers recognize that quality is an important attribute in products and services. Suppliers recognize that quality can be an important differentiator between

FIGURE 4.2   Continuous process of management of quality.

their own offerings and those of competitors. This quality differentiation is being called the "quality gap." In the past few decades these quality gaps have been greatly reduced between competitive products and services. The major reason has been the internationalization of trade and competition. These countries or companies have also raised their own standards of quality in order to meet international standards and customer demands. Quality consciousness, which started with manufacturing, has spread to the sales, marketing, and services sector as well. In recent times some themes have become more significant, including quality culture, the importance of knowledge management, and the role of leadership in promoting and achieving high quality. Disciplines like systems thinking are bringing more holistic approaches to quality so that people, processes, and products are considered together rather than as independent factors in quality management. Quality management in food is a way for managers, maintenance engineers, production supervisors, and others to ensure that any food product will satisfy the required standards fixed for that product. Apart from the product, it should also ensure that the procedures followed to develop, implement, and execute the product are efficient. Quality management consists of three main subcategories: quality planning, quality assurance, and quality control. Quality management is not a one-time process but a continuous and constant exercise. Figure 4.2 illustrates the process and cycle of quality management.

4.3.1   QUALITY PLANNING

In simple terms, quality planning is the task of identifying the standards by which you will measure the quality of a specific project, as well as identifying how to satisfy these standards. Several tools can be used for quality planning. A cost-benefit analysis is an effective tool to determine what the benefit of fewer defects, less reworking, and higher productivity will cost. This will apply to products, services, or other business processes. When you fully understand the costs associated with attaining a level of quality, the standards by which the quality of a project is measured can be selected (Rose, 2005).

4.3.2 QUALITY ASSURANCE

Quality assurance is the process by which we apply systematic quality activities like audits and checks and use feedback from these activities for continuous improvement. A traditional tool used in quality assurance is the quality audit. These audits are formal, structured reviews conducted independently to determine if the project is in compliance with the quality standards identified in the planning stage. Also, process analysis is often used as a quality assurance tool. Process analysis will, in many cases, include root cause analysis to help identify any underlying quality issues with specific components of a process, service, or product.

4.3.3 QUALITY CONTROL

Quality control is the task of monitoring all of the processes we have in place and being able to apply corrective action when anomalies are identified. Quality control has many tools and techniques; seven of these are known as the "Seven Basic Tools of Quality." They include: cause-and-effect diagrams (also known as Ishikawa or fishbone diagrams), control charts, flow charts, histograms, Pareto charts, run charts, and scatter diagrams.

It is inevitable that at some point you will need to deal with low quality in projects, products, or processes. Though good-quality planning will help overcome low quality, it is ultimately a marriage of all quality management concentrations and a good project manager that will have the greatest impact on quality. Quality should be the result of proper planning and ideal management. Quality improvement can be distinguished from quality control in that quality improvement is the purposeful change of a process to improve the reliability of achieving an outcome. Quality control is the ongoing effort to maintain the integrity of a process to maintain the reliability of achieving an outcome. Quality assurance is the set of planned or systematic actions necessary to provide enough confidence that a product or service will satisfy the given requirements.

4.4 PRINCIPLES OF QUALITY MANAGEMENT

There are certain guidelines or principles that are to be kept in mind before any strategy or system can be designed to manage quality in any organization. These principles will guide these organizations toward improved performance:

1. *Consumer satisfaction*: Since the organization caters to the consumers of their products, they are dependent on their customers. The organization has to understand current and future customer needs, should meet customer requirements, and should try to exceed the expectations of customers. An organization attains customer focus when all people in the organization work toward meeting customer requirements. The organization must ensure

that products conform to the expectations of the consumers, and that customers are satisfied.

2. *Commitment of leadership*: The top management of an organization must be committed to maintenance of quality. Quality objectives are to be established for the organization and its products. Leaders of an organization must establish unity and direction of purpose. They should go for the creation and maintenance of an internal environment in which people can become fully involved in achieving the organization's quality objective.

3. *Employee participation*: It is impossible to achieve the desired results without the active involvement and participation of all the people associated with the production and processing of food products. The people at all levels of an organization are the essence of it. Their complete involvement enables their abilities to be used for the benefit of the organization, even when the ultimate key decisions regarding quality issues are made by the project manager.

4. *Process approach*: Instead of the product approach, it is appropriate to have control and monitoring of the complete process. In the process control approach the focus is not lost on the product, but emphasis is placed on the process so that the product delivered is of the desired quality. The desired result can be achieved when activities and related resources are managed in an organization as a process.

5. *System approach to management*: This is an extension of the process approach. An organization's effectiveness and efficiency in achieving its quality objectives are contributed to by identifying, understanding, and managing all interrelated processes as a system. Like in process control, the process as a whole is important, along with the final product; the organization has to treat all the interrelated processes as a whole system because the failure of a part will result in the failure of the complete organization. Quality control involves checking transformed and transforming resources at all stages of the production process.

6. *Measures for continual improvement*: Improvement is a continuous process. Process performance measures are to be taken at all stages of production at all points of time. One of the permanent quality objectives of an organization should be the continual improvement of its overall performance, leveraging clear and concise process performance measures.

7. *Information collection and analysis*: Effective decisions are always based on data analysis and information. An organization must collect, collate, and analyze all relevant facts that are directly and indirectly related to and required for the production process.

8. *Raw material quality*: Suppliers are the main sources of quality raw material. If the quality of the raw material is good, the final product will be also of good quality. Since an organization and its suppliers are interdependent, a mutually beneficial relationship between them increases the ability of both to add value.

Figure 4.3 illustrates and summarizes the eight principles for quality management for any food business operation.

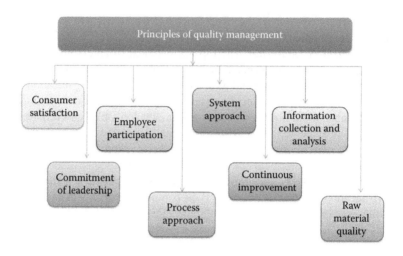

FIGURE 4.3   Principles of quality management.

4.5   GOOD MANUFACTURING PRACTICES AND STANDARD OPERATING PROCEDURES

Good manufacturing practices (GMPs) and standard operating procedures (SOPs) are two tools in, for example, a meat-processing facility that help in the production of high-quality and safe food products. The programs established for GMPs and SOPs will provide the basis for other programs to help assure the level of product quality, such as standards for ISO 9000 and for product safety in the Hazard Analysis and Critical Control Point (HACCP) system.

Current GMPs in food describe the methods, equipment, facilities, and controls in place for producing processed food. As the minimum sanitary and processing requirements for producing safe and wholesome food, they are an important part of regulatory control over the safety of the food supply. GMPs also serve as one basis for inspections by regulatory authorities. GMPs are the result of an experimentation and observation process that has spanned decades.

Food safety has been regulated since the mid-1800s and was mostly the responsibility of local and state regulators. The intent of covering GMPs in the regulations of most countries was to prevent contamination, poisoning, and fraud.

The general provisions of GMPs cover employee responsibilities with regard to personal hygiene. For example, personnel with diseases or other conditions that could contaminate food are to be excluded from manufacturing operations. GMPs also outline expectations with respect to personal hygiene and cleanliness, clothing, removal of jewelry and other unsecured objects, glove maintenance, use of hair restraints, appropriate storage of personal items, and restrictions on various activities such as eating and smoking. The general provisions of GMPs discuss the need for appropriate food safety education and training. Some subsections of food GMPs outline the requirements for the maintenance, layout, and operations of food processing facilities.

Further subsections outline the requirements for adequate maintenance of the grounds, including litter control, waste removal and treatment, and grounds maintenance and drainage. This subsection also requires that plants be designed and built to reduce the potential for contamination. Some detail is provided on how to achieve this, but the requirements are largely focused on the end result of a sanitary facility rather than specific practices. Physical facilities, equipment, and utensils are to be sanitized in a way that protects against food contamination. Storage of cleaning materials and permitted toxic materials are outlined to prevent contamination with chemicals. This subsection also briefly addresses pest control and the cleaning of various food contact surfaces, as well as the frequency of cleaning. It describes the requirements for adequate sanitary facilities and controls, including the water supply, plumbing, toilet and hand-washing facilities, and rubbish and offal disposal. Some of the requirements are fairly specific, such as the requirement of self-closing doors for toilet facilities, whereas others remain general, such as having plumbing of adequate size and design. There is a specific subsection that describes the requirements and expectations for the design, construction, and maintenance of equipment and utensils so as to ensure sanitary conditions. This also adds a specific requirement: an automatic control for regulating temperature, or an alarm system to alert employees to a significant change in temperature. Other requirements of this subsection are fairly general and intended to prevent contamination from any source. Production and process control is the most important part of GMPs; this lists the general sanitation processes and controls necessary to ensure that food is suitable for human consumption. This section also addresses the monitoring of physical factors (critical control points) such as time, temperature, humidity, pH, flow rate, and acidification. The next section of GMPs outlines the requirements for warehousing and distribution. It requires finished foods to be stored and distributed under conditions that protect against physical, chemical, and microbial contamination. The container and the food must also be protected from deterioration (FDA).

The maximum defect action levels for a defect that is natural or unavoidable, even when foods are produced under GMPs, is to be defined. Generally these defects are not hazardous to health at low levels; they include rodent filth, insects, or mold. The defect activation levels are defined for individual commodities.

4.6 TECHNIQUES AND METHODS OF QUALITY IMPROVEMENT

There are many methods of quality improvement. These cover product improvement, process improvement, system improvement/innovation, and people-based improvement. In the following sections are the methods and techniques of quality management that incorporate and drive quality improvement.

4.6.1 INTERNATIONAL ORGANIZATION FOR STANDARDIZATION

The International Organization for Standardization (ISO) created the quality management system (QMS) standards in 1987. They were the ISO 9000:1987 series of standards, comprising ISO 9001:1987, ISO 9002:1987, and ISO 9003:1987, which were applicable in different types of industries based on the type of activity or process: designing, production, or service delivery. These standards are reviewed every few years by the ISO. The 1994 version was called the ISO 9000:1994 series, consisting of the ISO 9001:1994, ISA 9002:1994, and ISO 9003:1994 versions. The last major revision was in 2008 and was called the ISO 9000:2000 series. The ISO 9002 and 9003 standards were integrated into one single certifiable standard: ISO 9001:2000. After December 2003, organizations holding ISO 9002 or 9003 standards had to complete a transition to the new standard. ISO released a minor revision, ISO 9001:2008, on October 14, 2008 (Cianfrani and West, 2009; Westcott, 2003). It contained no new requirements; many of the changes were to improve consistency in grammar, facilitating translation of the standard into other languages for use by over 950,000 certified organizations in the 175 countries that use the standard (as of December 2007). The ISO 9004:2009 document gives guidelines for performance improvement over and above the basic standard (ISO 9001:2000). This standard provides a measurement framework for improved quality management similar to and based upon the measurement framework for process assessment. The QMS standards created by ISO are meant to certify the processes and systems of an organization, not the product or service itself. ISO 9000 standards do not certify the quality of the product or service.

4.6.1.1 ISO 9004:2009

As mentioned above, the main purpose of ISO 9004 is to provide a guide for performance. The 9000 series QMS consists of three standards: ISO 9000, which covers fundamentals; ISO 9001, which covers requirements; and ISO 9004, which covers performance improvements. The last version of this QMS occurred in 2009, hence the term "ISO 9004:2009" is used. This international standard provides guidelines beyond the requirements given in ISO 9001 in order to consider both the effectiveness and efficiency of a QMS, and consequently the potential for improvement of the performance of an organization. When compared to ISO 9001, the objectives of customer satisfaction and product quality are extended to include the satisfaction of interested parties and the performance of the organization. This international standard is applicable to the processes of the organization, and consequently the quality management principles on which it is based can be deployed throughout the organization. The focus of this international standard is the achievement of ongoing improvement measured through the satisfaction of customers and other interested parties. This international standard consists of guidance and recommendations and is not intended for certification, regulatory, or contractual use, nor as a guide to the implementation of ISO 9001 (Indian Register Quality Systems).

In 2005 ISO released a standard, ISO 22000, meant for the food industry. This standard covers the values and principles of ISO 9000 and the HACCP standards. It gives one single integrated standard for the food industry and is expected to become more popular in the industry in the coming years.

ISO has also released standards for other industries. For example, technical standard TS 16949 defines requirements in addition to those in ISO 9001:2008 specifically for the automotive industry. ISO has a number of standards that support quality management. One group (including ISO/IEC 12207 and ISO/IEC 15288) describes processes, and another (ISO 15504) describes process assessment and improvement.

4.6.1.2   ISO/IEC 15504 and ISO/IEC 33001:2015

ISO/IEC 15504 and ISO/IEC 33001:2015 are also termed the "Software Process Improvement and Capability dEtermination." This is a set of technical standards documents for the computer software development process and related business management functions. It is one of the joint ISO and International Electrotechnical Commission (IEC) standards, which was developed by the ISO and IEC joint subcommittee, ISO/IEC JTC 1/SC 7. ISO/IEC 15504 was initially derived from process lifecycle standard ISO/IEC 12207 and from maturity models like Bootstrap, Trillium, and the Capability Maturity Model (CMM). ISO/IEC 15504 was revised by ISO/IEC 33001:2015 as of March 2015 and is no longer available at ISO.

4.6.2   QUALITY FUNCTION DEPLOYMENT

Quality function deployment (QFD) is referred to by many names, including "matrix product planning," "decision matrices," and "customer-driven engineering." QFD is a focused methodology for carefully listening to the voice of the customer and then effectively responding to those needs and expectations. First developed in Japan in the late 1960s as a form of cause-and-effect analysis, QFD was brought to the United States in the early 1980s. It gained its early popularity as a result of numerous successes in the automotive industry. In QFD, quality is a measure of customer satisfaction with a product or service. QFD is a structured method that uses the seven management and planning tools to identify and prioritize customers' expectations quickly and effectively. Beginning with the initial matrix, commonly termed the "house of quality," the QFD methodology focuses on the most important product or service attributes or qualities. These are composed of customer wows, wants, and musts. Once we have prioritized the attributes and qualities, QFD deploys them to the appropriate organizational function for action. Thus, QFD is the deployment of customer-driven qualities to the responsible functions of an organization. Many QFD practitioners claim that using QFD has enabled them to reduce their product and service development cycle times by as much as 75%, with equally impressive improvements in measured customer satisfaction.

4.6.3   KAIZEN

"Kaizen" is a Japanese word that means the practice of continuous improvement. Kaizen was originally introduced to the West by Masaaki Imai in his book *Kaizen: The Key to Japan's Competitive Success* (1986). Today kaizen is recognized worldwide as an important pillar of an organization's long-term competitive strategy. Kaizen is continuous improvement that is based on certain guiding principles:

- Good processes bring good results.
- You must make assessments yourself to grasp the current situation.
- Speak with data, manage by facts.

- Take action to contain and correct root causes of problems.
- Work as a team.
- Kaizen is everybody's business.
- Big results come from many small changes accumulated over time.

Kaizen is teamwork, which means that everyone is involved in making improvements. While the majority of changes may be small, the greatest impact may be kaizens that are led by senior management as transformational projects, or by cross-functional teams as kaizen events.

4.6.4 ZERO DEFECT PROGRAM

The phrase "zero defects" was coined by Philip Crosby in his 1979 book *Quality is Free*. His position was that, where there are zero defects, there are no costs associated with issues of poor quality, and hence quality becomes free. Zero defects in quality management cannot be taken literally as it is not possible to have zero defects. Instead, it refers to a state in which waste is eliminated and defects are reduced. It means ensuring quality standards and reducing defects to the level of zero in projects. Zero defects is a concept of the "quest for perfection" in order to improve quality. Though perfection might not be achievable, at least the quest will lead toward improvements in quality. The zero defects theory also closely connects with the phrase "right first time." This means that every project should be perfect the very first time. Here, again, "perfect" refers to zero defects. The zero defects theory is based on four elements for implementation in real projects:

1. *Quality is a state of assurance to requirements*: Zero defects in a project means fulfilling requirements at that point of time.
2. *Right first time*: Quality should be taken care of at the very first go, rather than solving problems at a later stage.
3. *Quality is measured in financial terms*: One needs to judge waste, production, and revenue in terms of money.
4. *Performance should be judged by the zero defects theory*: That is, near to perfection. Just being good is not good enough.

4.6.5 SIX SIGMA

The term "Six Sigma" originated from terminology associated with the statistical modeling of manufacturing processes. Six Sigma (6σ) combines established methods such as statistical process control, design of experiments, and failure mode and effects analysis (FMEA) in an overall framework. Six Sigma is a set of techniques and tools for process improvement. It was introduced by engineer Bill Smith while working at Motorola in 1986. Jack Welch made it central to his business strategy at General Electric in 1995. Today it is used in many industrial sectors. Six Sigma seeks to improve the quality of the output of a process by identifying and removing the causes of defects and minimizing variability in manufacturing and business processes. It uses a set of quality management methods—mainly empirical, statistical methods—and creates a special infrastructure of people within the organization who are experts in these methods. Each Six Sigma project carried out within an organization follows a defined sequence of steps and has specific value targets—for example, to reduce process cycle time, pollution, and costs and increase customer satisfaction and profits.

4.6.6 PDCA

The "plan–do–check–act" cycle (PDCA) is also sometimes called the "plan–do–study–act" cycle (PDSA), the "Deming cycle," or the "Shewhart cycle." PDCA was made popular by Dr. W. Edwards Deming, who is considered by many to be the father of modern quality control; he always referred to it as the Shewhart cycle. Later in Deming's career, he modified PDCA to PDSA because he felt that the term "check" emphasized inspection over analysis. The PDCA cycle is a four-step model for carrying out change. Just as a circle has no end, the PDCA cycle should be repeated again and again for continuous improvement. The steps in this are

1. *Plan*: Recognize an opportunity and plan a change.
2. *Do*: Test the change. Carry out a small-scale study.
3. *Check*: Review the test, analyze the results, and identify what you've learned.
4. *Act*: Take action based on what you learned in the "check" step. If the change did not work, go through the cycle again with a different plan. If you were successful, incorporate what you learned from the test into wider changes. Use what you learned to plan new improvements, beginning the cycle again.

In Six Sigma programs, the PDCA cycle is called "define, measure, analyze, improve, control" (DMAIC). The iterative nature of the cycle must be explicitly added to the DMAIC procedure.

4.6.7 QUALITY CIRCLE

A quality circle is a participatory management technique that enlists the help of employees in solving problems related to their jobs. Circles of employees working together in an operation are formed, and they meet at intervals to discuss problems of quality and to devise solutions for improvements. Quality circles have an autonomous character, are usually small, and are led by a supervisor or a senior worker. Employees who participate in quality circles usually receive training in formal problem-solving methods such as brainstorming, Pareto analysis, and cause-and-effect diagrams and are then encouraged to apply these methods either to specific or general company problems. After completing an analysis, they often present their findings to management and then handle the implementation of approved solutions. The principle of quality circles has simply moved quality control to an earlier position in the production process. Rather than relying upon postproduction inspections to catch errors and defects, quality circles attempt to prevent defects from occurring in the first place. As an added bonus, machine downtime and the accumulation of scrap materials that formerly occurred due to product defects are minimized. Deming's idea that improving quality could increase productivity led to the development in Japan of the total quality control concept, in which quality and productivity are viewed as two sides of a coin. Total quality control also requires that a manufacturer's suppliers make use of quality circles (Thareja, 2008).

4.6.8 TAGUCHI METHODS

Taguchi methods are part of a quality control methodology that combines control charts and process control with product and process design to achieve a robust total design. The methodology aims to reduce product variability with a system for developing specifications and designing them into a product or process. It is named after its inventor, the Japanese engineer-statistician Dr. Genichi Taguchi, who also developed the quality loss function.

4.6.9 LEAN MANUFACTURING

Lean manufacturing or lean production (often simply called "lean") is a systematic method for the elimination of waste ("Muda") within a manufacturing system. Lean also takes into account waste created through overburden ("Muri") and waste created through unevenness in workloads ("Mura"). Working from the perspective of the client who consumes a product or service, "value" is any action or process for which a customer would be willing to pay. Essentially, lean is centered on making obvious what adds value, by reducing everything else. Lean manufacturing is a management philosophy derived mostly from the Toyota production system (hence the term "Toyotism" is also prevalent) and was identified as "lean" only in the 1990s. The Toyota production system is renowned for its focus on the reduction of the original Toyota seven wastes to improve overall customer value, but there are varying perspectives on how this is best achieved. The steady growth of Toyota from a small company to the world's largest automaker has focused attention on how it has achieved this success.

4.6.10 KANSEI ENGINEERING

Kansei engineering is a method for translating feelings and impressions into product parameters. The method was invented in the 1970s by Professor Nagamachi at Kure University (now Hiroshima International University). Professor Nagamachi recognized that companies often want to quantify the customer's impression of their products. Kansei engineering can "measure" these feelings and show their correlation to certain product properties. As a consequence, products can be designed in such a way that they respond to the intended feeling. This is a multidisciplinary area in which experts from different fields give their feedback, ideas, and improvement methods. Quantitative data based on the feedback of customers is collected and then analyzed statistically so as to link the physical properties of the product to the elicited perceptions. Kansei engineering aims to develop or improve products and services by translating the customer's psychological feelings and needs into the domain of product design (i.e., the parameters).

4.6.11 TOTAL QUALITY MANAGEMENT

Total quality management (TQM) is a comprehensive and structured approach to organizational management that seeks to improve the quality of products and services through ongoing refinements in response to continuous feedback. TQM is a management strategy aimed at embedding the awareness of quality in all organizational processes. It was first promoted in Japan with the Deming Prize, which was adopted and adapted in the United States as the Malcolm Baldrige National Quality Award and in Europe as the European Foundation for Quality Management Award (each with its own variations). TQM is a management philosophy that seeks to integrate all organizational functions (e.g., marketing, finance, design, engineering and production, customer service, etc.) to focus on meeting customer needs and organizational objectives (Omachonu and Ross, 2004).

TQM views an organization as a collection of processes. It maintains that organizations must strive continuously to improve these processes by incorporating the knowledge and experiences of workers. The simple objective of TQM is: "Do the right things, right the first time, every time." TQM is infinitely variable and adaptable. Although originally applied to manufacturing operations, and for a number of years only used in that area, TQM is now becoming recognized as a generic management tool just as applicable to service and public sector organizations. There are a number of evolutionary strands to TQM, with different sectors creating their own versions from the common ancestor. TQM is the foundation for activities that include

- Commitment by senior management and all employees
- Meeting customer requirements
- Reducing development cycle times
- Just-in-time/demand flow manufacturing
- Improvement teams
- Reducing product and service costs
- Creating systems to facilitate improvement
- Line management ownership
- Employee involvement and empowerment
- Recognition and celebration
- Challenging quantified goals and benchmarking
- Focus on processes/improvement plans
- Specific incorporation in strategic planning

This list shows that TQM must be practiced in all activities, by all personnel in manufacturing, marketing, engineering, research and development, sales, purchasing, human resources, and so on (Hyde, 1992).

4.6.12 THEORY OF INVENTIVE PROBLEM SOLVING

"TRIZ" is the (Russian) acronym for the "theory of inventive problem solving." TRIZ is a problem-solving method based on logic and data, not intuition, which accelerates the project team's ability to solve problems creatively. TRIZ also provides repeatability, predictability, and reliability due to its structure and algorithmic approach. G.S. Altshuller and his colleagues in the former U.S.S.R. developed the method between 1946 and 1985. TRIZ is an international science of creativity that relies on the study of the patterns of problems and solutions, not on the spontaneous and intuitive creativity of individuals or groups. More than three million patents have been analyzed to discover the patterns that predict breakthrough solutions to problems. TRIZ is increasingly common in Six Sigma processes, in project management and risk management systems, and in organizational innovation initiatives. TRIZ research began with the hypothesis that there are universal principles of creativity that are the basis for creative innovations that advance technology. If these principles could be identified and codified, they could be taught to people to make the process of creativity more predictable. The research has proceeded in several stages during the last 60 years. The three primary findings of this research are as follows:

1. Problems and solutions are repeated across industries and sciences; the classification of the contradictions in each problem predicts the creative solutions to that problem.
2. Patterns of technical evolution are repeated across industries and sciences.
3. Creative innovations use scientific effects outside the field in which they were developed.

Much of the practice of TRIZ consists of learning these repeating patterns of problems–solutions, patterns of technical evolution, and methods of using scientific effects, and then applying the general TRIZ patterns to the specific situation that confronts the developer.

4.6.13 BUSINESS PROCESS REENGINEERING

The concept of business process reengineering (BPR) was first introduced in the late Michael Hammer's 1990 *Harvard Business Review* article: "Reengineering Work: Don't Automate, Obliterate," and received increased attention a few years later when Hammer and James Champy published their bestselling book, *Reengineering the Corporation*. The authors promoted the idea that sometimes the radical redesign and reorganization of an enterprise is necessary to lower costs and increase quality of service, and that information technology is the key enabler for that radical change. Hammer and Champy suggested seven reengineering principles to streamline the work process and thereby achieve significant levels of improvement in quality, time management, speed, and profitability:

1. Organize around outcomes, not tasks.
2. Identify all the processes in an organization and prioritize them in order of redesign urgency.
3. Integrate information processing work into the real work that produces the information.
4. Treat geographically dispersed resources as though they were centralized.
5. Link parallel activities in the workflow instead of just integrating their results.

6. Put the decision point where the work is performed, and build control into the process.
7. Capture information once and at the source.

BPR involves the radical redesign of core business processes to achieve dramatic improvements in productivity, cycle times, and quality. In BPR, companies start with a blank sheet of paper and rethink existing processes in order to deliver more value to the customer. They typically adopt a new value system that places increased emphasis on customer needs. Companies reduce organizational layers and eliminate unproductive activities in two key areas. First, they redesign functional organizations into cross-functional teams. Second, they use technology to improve data dissemination and decision making.

4.6.14 VDA 6.1

VDA 6.1 is a German QMS standard created by the German Automotive Industry Association (VDA) and initiated by the automobile industry. The first VDA standard was for the exchange of surface models and was named "VDA-FS." It has been superseded by a subset of the "Initial Graphics Exchange Specification" (IGES), referred to simply as "VDA." Aside from this exchange standard, VDA also developed "VDA-PS," a library standard for standard parts, now known as "DIN 66304." Based on ISO 9001:1994, the QMS includes all elements of QS-9000, with an additional four requirements specific to VDA 6.1, as follows:

1. *Element 06.3 on "Recognition of Product Risk"*: This is the risk of the product failing to fulfill its stipulated function, and its effect on the whole assembly.
2. *Element Z1.5 on "Employee Satisfaction"*: This covers the perception of the company's employees, as well as their needs and expectations, which will be met through the company's quality system.
3. *Element 07.3 on "Quotation Structure"*: A customer or market is offered products for purchase or that are made available to own or to use.
4. *Element 12.4 on "Quality History"*: This section covers the quality history of customer-supplied products and gives an overview of the situation during a particular period.

4.6.15 OBJECT-ORIENTED QUALITY AND RISK MANAGEMENT

Object-oriented quality and risk management is a model that integrates quality and risk management. It targets the innovations focused on quality improvements (Van Nederpelt, 2012). It aims to achieve the following:

- Innovation in the domain of quality and risk management
- Applicability in any organization, on any scale, and in any field of expertise
- Customized frameworks, quality assurance, and risk management
- Integrated quality and risk management
- Practical and efficient processes
- A standard approach for customized solutions

4.6.16 Continuous Quality Improvement Process

The continuous quality improvement process (CQI) uses a team approach to accomplish operational changes. Change occurs by following sequential steps that focus on changing procedures, empowering employees, placing customers first, and achieving long-term organizational commitment. The CQI process uses six sequential steps referred to as the "problem-solving discipline approach" (Rampersad, 2001):

1. Define area(s) for improvement.
2. Identify all possible causes.
3. Develop a CQI action plan.
4. Implement the CQI action plan.
5. Evaluate measurement outcomes for program improvement.
6. Standardize the CQI process.

The success of these techniques depends on the commitment, knowledge, and expertise to guide improvement, scope of change/improvement desired, and adaption to enterprise cultures. Big and comprehensive changes tend to fail more often compared to smaller changes. Selection of a particular technique in an organization is important: quality circles do not work well in every enterprise, and there have been well-publicized failures of BPR, as well as Six Sigma. Therefore, enterprises need to carefully consider which quality improvement methods to adopt, and certainly should not adopt all those listed here. It is important not to underestimate the human factors such as culture in selecting a quality improvement approach. Any improvement (i.e., change) takes time to implement, gain acceptance, and stabilize as accepted practice. Improvement must allow for pauses between implementing new changes, so that the change can be stabilized and assessed as a real improvement before the next improvement is made. Hence continual improvement is not continuous, but efforts must continue after achieving one step and then going on to the next one. Improvements that change the culture may take longer as it takes time to change habits, and normally in these circumstances there is greater resistance to change. It is easier and often more effective to work within the existing cultural boundaries and make small improvements than it is to make major transformational changes. The use of kaizen in Japan is considered to be a major reason for the creation of Japanese industrial and economic strength. On the other hand, transformational changes work best when an enterprise faces a crisis and needs to make major changes in order to survive. Well-organized quality improvement programs take all these factors into account when selecting quality improvement methods/techniques.

4.6.17 Quality Management Software

Quality management software is a category of technologies used by organizations to manage the delivery of high-quality products. Solutions range in functionality; however, with the use of automation capabilities they typically have components for managing internal and external risk, compliance, and the quality of processes and products.

Preconfigured and industry-specific solutions are available and generally require integration with existing IT architecture applications such as ERP, SCM, CRM, and PLM. Specific software has been developed to address the specific issues and facilitate the monitoring and assessment of quality parameters. These are used for

- Nonconformances/corrective and preventive action
- Compliance/audit management
- Supplier quality management
- Risk management
- Statistical process control
- Failure mode and effects analysis
- Complaint handling
- Advanced product quality planning
- Environment, health, and safety
- The HACCP approach
- The production part approval process

The intersection of technology and quality management software prompted the emergence of a new software category: enterprise quality management software (EQMS). EQMS is a platform for cross-functional communication and collaboration that centralizes, standardizes, and streamlines quality management data from across the value chain. The software breaks down functional silos created by traditionally implemented stand-alone and targeted solutions. Supporting the proliferation and accessibility of information across supply chain activities, design, production, distribution, and service, it provides a holistic viewpoint for managing the quality of products and processes (Loon, 2007).

A QMS is a collection of business processes focused on achieving quality policy and quality objectives to meet customer requirements. It is expressed as the organizational structure, policies, procedures, processes, and resources needed to implement quality management. Early systems emphasized predictable outcomes of an industrial product production line using simple statistics and random sampling. By the twentieth century, labour inputs were typically the most costly inputs in most industrialized societies, so focus shifted to team cooperation and dynamics, especially the early signaling of problems via a continuous improvement cycle. In the twenty-first century, QMS has tended to converge with sustainability and transparency initiatives, as both investor and customer satisfaction and perceived quality are increasingly tied to these factors. Of all QMS regimes, the ISO 9000 family of standards is probably the most widely implemented worldwide—the ISO 19011 audit regime applies to both, and deals with quality and sustainability and their integration (Littlefield and Roberts, 2012).

The Software Engineering Institute has its own process assessment and improvement methods called Capability Maturity Model Integration (CMMI) and Initiating, Diagnosing, Establishing, Acting and Learning (IDEAL), respectively. CMMI is a process improvement training and appraisal program and service administered and marketed by Carnegie Mellon University and required by U.S. government contracts, especially in software development. Carnegie Mellon University claims CMMI can be used to guide process improvement across a project, division, or entire organization. Under the CMMI methodology, processes are rated according to their maturity levels, which are defined as: initial, managed, defined, quantitatively managed, and optimizing. CMMI Version 1.3 (released on November 1, 2010) is currently supported, and CMMI is registered in the U.S. Patent and Trademark Office by Carnegie Mellon University. Three models of CMMI are: product and service development ("CMMI for Development"), service establishment, management, and delivery ("CMMI for Services"), and product and service acquisition ("CMMI for Acquisition"). The release of CMMI Version 1.3 is noteworthy because it updated all three of these CMMI models to make them consistent and to improve their high-maturity practices (CMMI).

4.7   ACCREDITATION OF LABORATORIES

Accreditation of food safety laboratories has become necessary to establish confidence in consumers that the product being used is safe and conforms to the quality as declared on the packaging. Accreditation of such laboratories involves a process of trusting the job to food testing laboratories to ensure that the biological and chemical components of the food any organization processes or packages are safe for consumers. Food business operators, as well as regulatory agencies around the world, are adopting accreditation processes so that there is uniformity and confidence in quality. Using an accredited laboratory to assess safety and quality also gives confidence to government bodies and regulators that the standards and regulations they set are being followed. In this process a third party, an accreditation body, assesses a testing lab's competency, as well as its compliance with relevant standards. The accreditation body will examine the proficiency of laboratory staff, verify accredited scopes of testing and calibration, and monitor the effectiveness of QMSs. Once a laboratory is accredited for certain tests or criteria, it is qualified to conduct tests and assess the quality of food products, with results that are accepted by everyone in the world. This enhances credibility and ensures uniformity of standards, as well as contributing to a higher level of safety around the world. Accreditation also certifies that a lab operates to a global standard and has passed a rigorous examination of its methods, facilities, and staff. This enhances credibility that the testing laboratory is producing data that is accurate, traceable, and reproducible (GFSR).

The National Association of Testing Authorities (NATA) is the world's first and most experienced laboratory accreditation body. Since its creation in 1947, it has been a founding member of the International Laboratory Accreditation Cooperation (ILAC) and the Asia Pacific Laboratory Accreditation Cooperation (APLAC), and provides both of their secretariats. NATA was born of necessity in the Second World War when Australia was cut off from any means of ensuring that the munitions it was manufacturing were of a sufficiently high standard. The notion of ensuring that testing standards were themselves subject to examination was then a novel one. In late 1945, a conference on the coordination of testing services, attended by representatives of all state and federal governments, led to the formation of NATA a little more than a year later. The new association was to provide a national testing service to Australia and would span all technical, industrial, and geographical areas of the country.

4.8 PLANNING FOR QUALITY AND QUALITY MANAGEMENT PLAN

Planning for food quality is a very important and systematic procedure, as quality can be achieved by carefully designating the processes and procedures. Organizations want to produce quality food that customers can appreciate and that can also establish brand value. It is a simple process to identify and learn what standards or quality parameters are necessary for the food and how to meet them. Quality planning is the task of identifying the determinants of food quality for a particular product and organizing how to fulfill these requirements. Such determinants often include the resources that will be used, the steps needed to produce the food, and any other specifications/standards. Thus it is a systematic process that translates quality policy into measurable objectives and requirements, and lays down a sequence of steps for realizing them within a specified timeframe.

4.8.1 QUALITY MANAGEMENT PLAN

The quality management plan is a key output of an organization's quality planning, which explains in detail the implementation plan to achieve the laid-down quality policy. It includes the steps that will be taken to address the issues of quality control, quality assurance, and continuous process improvement.

A quality management plan typically covers the following issues:

1. *Quality objectives and goals*: Every organization is required to identify the overall objectives of the quality management program, together with measurable goals to be achieved. Deliverables are to be fixed that should be realistic and achievable.
2. *Organizational structure and distribution of responsibilities*: Organizational structure, reporting relationships, and roles and responsibilities for quality must be clearly defined and explained to all the concerned officials so that there is no confusion.
3. *Resource requirements*: Identification of the people and resources required to develop, use, maintain, and improve the QMS is one of the most essential components of the plan.
4. *Quality checklists*: A checklist is a structured tool, usually in the form of a table, listing all the essential steps/stages. It is normally component-specific and is used to verify that a set of required steps has been performed. These act as guidance papers as well. Checklists range from simple to quite complex and are designed as per the requirements.
5. *Process improvement plan*: The process improvement plan details the steps to be employed in analyzing program processes, with a view to identifying wasteful, redundant, non-value-added activities. It typically consists of a detailed plan and chart showing the process flow from start to end so that areas with scope for improvement can be analyzed. Based on this analysis, targets for improved performance are fixed.

6. *Activities and deliverables*: A list of activities, process flow, and deliverables is prepared so that progress can be monitored.
7. *Quality metrics*: A quality metric explains how the quality control/quality assurance process will measure quality. These are the metrics of time line and deliverables, which sets the time frames in which work will be achieved, together with major milestones for QMS element delivery, review, and deployment.
8. *Quality baseline*: The quality baseline records the program's established quality objectives and quality metrics. A program measures and records its quality performance with reference to the quality baseline.
9. *Risk analysis*: An analysis of what could go wrong, together with strategies for risk reduction.

Formulation of a detailed plan for addressing the issue of quality in any food product is very important for the success of any program and activity. This gives an opportunity to identify/fix quality parameters, identify deliverables, fix timelines, and cater for the deviations or likely challenges to achieve the desired quality.

APPENDIX 4.A

ILAC: Accreditation is the independent evaluation of conformity assessment bodies against recognized standards to carry out specific activities to ensure their impartiality and competence. Through the application of national and international standards, government, procurers and consumers can have confidence in the calibration and test results, inspection reports and certifications provided. Accreditation bodies are established in many economies with the primary purpose of ensuring that conformity assessment bodies are subject to oversight by an authoritative body. Accreditation bodies, that have been peer evaluated as competent, sign regional and international arrangements to demonstrate their competence. These accreditation bodies then assess and accredit conformity assessment bodies to the relevant standards. The arrangements support the provision of local or national services, such as providing safe food and clean drinking water, providing energy, delivering health and social care, or maintaining an unpolluted environment. In addition, the arrangements enhance the acceptance of products and services across national borders, thereby creating a framework to support international trade through the removal of technical barriers.

The international arrangements are managed by ILAC, in the fields of calibration, testing, medical testing and inspection accreditation, and the International Accreditation Forum (IAF), in the fields of management systems, products, services, personnel and other similar programmes of conformity assessment. Both organizations, ILAC and IAF, work together and coordinate their efforts to enhance the accreditation and the conformity assessment worldwide. The regional arrangements are managed by the recognized regional co-operation bodies that work in harmony with ILAC and IAF. The recognized regional co-operations are also represented on the ILAC and IAF Executive Committees. ILAC works closely with the regional co-operation bodies involved in accreditation, notably EA in Europe, APLAC in the Asia-Pacific, IAAC in the Americas, AFRAC in Africa, SADCA in Southern

Africa, and ARAC in the Arab region. ILAC is the international organization for accreditation bodies operating in accordance with ISO/IEC 17011 and involved in the accreditation of conformity assessment bodies including calibration laboratories (using ISO/IEC 17025), testing laboratories (using ISO/IEC 17025), medical testing laboratories (using ISO 15189), and inspection bodies (using ISO/IEC 17020).

APPENDIX 4.B

IAS Food Safety Related Accreditation Programs: The International Accreditation Service (IAS), an internationally recognized nonprofit organization based in Southern California, provides accreditation in the food service industry. The IAS accreditation programs and services listed below encompass the diverse needs required within the food industry to enable farmers, manufacturers, distributors, and service providers to provide consumers with safe and quality consumable products:

- Testing Laboratories: Laboratories can be accredited to ISO/IEC Standard 17025: 2005, General Requirements for the Competence of Testing and Calibration Laboratories, and related standards (USDA, Codex, AOAC, USFDABAM, EPA, etc.)
- Calibration Laboratories: Calibration providers can be accredited to ISO/IEC Standard 17025:2005, General Requirements for the Competence of Testing and Calibration Laboratories, and related standards.

REFERENCES

Cianfrani C.A. and West J.E. (2009). *Cracking the Case of ISO 9001:2008 for Service: A Simple Guide to Implementing Quality Management to Service Organizations*, 2nd edn. Milwaukee, Wisconsin: American Society for Quality.
CMMI. (2007). Capability Maturity Model Integration version 1.2 overview. Carnegie Mellon University. www.cmmiinstitute.com. Accessed on February 9, 2016.
FDA. (2005). Food CGMP modernization—A focus on food safety. http://www.fda.gov/Food/Guidance Regulation/CGMP/default.htm. Accessed on February 9, 2016.
GFSR. http://globalfoodsafetyresource.com/food-science/food-laboratory-accreditation. Accessed on October 12, 2015.
Hyde A. (1992). The proverbs of total quality management: Recharting the path to quality improvement in the public sector. *Public Productivity and Management Review* 16(1): 25–37.
IAS. (2014). Accreditation in the food supply chain. http://www.iasonline.org/More/2014/Food_Safety_Brochure.pdf. Accessed on October 12, 2015.
ILAC. http://ilac.org/about-ilac/. Accessed on October 12, 2015.
Indian Register Quality Systems. (2015). ISO 9001 Quality Management System QMS Certification. www.irqs.co.in. Accessed on February 9, 2016.
Littlefield M. and Roberts M. (June 2012). Enterprise quality management software best practices guide. *LNS Research Quality Management Systems*. 10: 21–24.
Loon H.W. (2007). *Process Assessment and Improvement: A Practical Guide*, 2nd edn. Springer.
NATA. http://www.nata.com.au/nata/about-nata/history-of-nata. Accessed on October 12, 2015.
Omachonu V.K. and Ross J.E. (2004). *Principles of Total Quality*, 3rd edn. CRC Press.
PDCA. Taking the First Step with PDCA. February 2, 2009. Retrieved March 17, 2016.

Rose K.H. (July 2005). *Project Quality Management: Why, What and How*. Fort Lauderdale, FL: J. Ross Publishing.

Selden PH. (December 1998). Sales process engineering: An emerging quality application. *Quality Progress* 31(12): 59–63. Quality Management Strategy, May 2010.

TELARC. http://www.telarc.co.nz/food-safety/. Accessed on October 12, 2015.

Thareja P. (July/August 2008). A total quality organization thro' people: Part 16—Each one of is Capable, *FOUNDRY. Journal For Progressive Metal Casters* 20(4): 13.

The Economist. (2013). A healthy future for all? Improving food quality for Asia. A Report by The Economist Intelligence Unit Limited. www.eiuperspectives.economist.com/sites/default/files/A%20Healthy%20Future%20For%20All_main_Oct16_V3.pdf, Accessed on October 22, 2015.

United Nations. (2011). The Millennium Development Goals Report. www.un.org/millenniumgoals/pdf/(2011_E)%20MDG%20Report%202011_Book%20LR.pdf, Accessed on October 22, 2015.

Van Nederpelt P.W.M. EMEA. (2009). Object-Oriented Quality Management, a model for quality management. Statistics Netherlands, The Hague, the Netherlands.

Van Nederpelt P.W.M. (2012). *Object-oriented Quality and Risk Management (OQRM). A Practical and Generic Method to Manage Quality and Risk*. Alphen aan den Rijn, the Netherlands: Micro Data BV, October 24, 2012.

Westcott RT. (2003). *Stepping up to ISO 9004:2000: A Practical Guide for Creating a World-Class Organization*. Chico, CA: Paton Press LLC.

5 Management of Food Safety

5.1 CHALLENGES IN FOOD SAFETY MANAGEMENT

The responsibility of providing safe food to the consumer is shared by all the people involved at every stage in the production of food. Since most consumers receive their food from retail and food service establishments, a significant share of the responsibility for providing safe food to the consumer rests with these facilities. Operators of retail and food service establishments can make the greatest impact on food safety. The common goal of operators and regulators of retail and food service establishments is to produce safe, quality food for consumers. Operators of retail and food service establishments routinely respond to inspection findings by correcting violations, but they often do not implement proactive systems of control to prevent violations from recurring. While inspections and enforcement systems contribute a great deal to improving basic sanitation and upgrading facilities, they emphasize reactive rather than preventive measures to food safety. Additional measures must be taken on the part of operators and regulators to better prevent or reduce foodborne illnesses. Safe and adequate food is a human right, safety being a prime quality attribute without which food is unfit for consumption. Food safety regulations are framed to exercise control over all types of food produced, processed, and sold so that the customer is assured that the food consumed will not cause any harm. Every country has been taking adequate measures through practices, codes, regulations, and legislation to ensure that food available in the market is safe. Industrialization resulted in changes in lifestyles and also the production and processing of food in factories on a large scale. Developed countries like the United States and the United Kingdom were the first to respond to the changes in the lifestyles and demands of the consumers. From the perspective of emerging and developing countries, implementation of food regulations is required to improve food and nutrition security, the food trade, and the delivery of safe, ready-to-eat (RTE) foods at all places and at all times. The Millennium Development Goals (MDGs) put forward to transform developing societies include goals that translate to food safety. The success of the MDGs, including that of poverty reduction, will in part depend on an effective reduction of foodborne diseases, particularly among the vulnerable, which include women and children. Food- and waterborne illnesses can be serious health hazards and are responsible for high incidences of morbidity and mortality across all age groups. The enforcement of food regulations would assist in facilitating food trade within and outside developing countries through better compliance, ensuring the safety of RTE catered foods, as well as addressing issues related to the environment. At the same time, regulations need to be optimal, as overregulation may have undue negative effects on the food trade. In most of the countries, food safety has gained importance and legal frameworks for food safety are in place. The control institutions have been mandated, and regulations and standards have been implemented along local and regional food supply chains. Still, there is a requirement to have a robust national food safety control system in place. The main reasons why a few countries are not able to have these systems are as follows:

- Consumers' lack of awareness about food safety issues in conjunction with low income levels, leading them to make buying decisions based on prices and not on (largely invisible) food safety and quality criteria
- Limited capacities of the national institutions responsible for controlling compliance with plant protection, animal health, and food safety provisions to adapt and adopt international or regional standards
- Lack of awareness about food safety and quality issues and the lack of incentives for food supply chain operators—from inputs, through farming, trading, and processing, up to retail—to invest in good practices and quality assurance systems and further also to the basic product quality attribute such as visual features

A complete and comprehensive process has to be followed for ensuring that consumers are able to get safe food. The quality of food is also equally important. Safety of food can be ensured by adopting processes and procedures that will set off an alarm whenever safety has been compromised. However, quality has wider connotations and interpretations, which may involve issues of food security and nutrition as well. To have all such issues addressed in any organization, the solution lies in the establishment of a food safety management program/system. Any such system or program will begin with the systematic identification of food safety hazards that are reasonably likely to occur. During the food handling operations of the business, there are a number of occasions when food can be made unsafe and unsuitable for consumption. The business systematically examines its food handling operations by listing the steps used to produce food in the business in a logical, progressive sequence—that is, from the receipt of food until its final step for sale. A food safety hazard may be a biological, chemical, or physical agent in or a condition of the food that creates the potential for it to be unsafe or unsuitable and cause an adverse health effect in humans. It is only necessary for hazards to be identified if these are reasonably likely to occur due to the specific nature, storage, transportation, preparation, or handling of the food. It is not reasonable to expect businesses to identify hazards that have not yet been discovered, such as new harmful bacteria. The management and control of such identified hazards follows. Every successful food business has to ensure that it is adopting a method or process through which the identified hazard can be controlled. Hazards may be controlled by support programs or at the specific food handling steps. These controls (alone or collectively) must be effective in preventing, eliminating, or reducing the hazard to a safe, acceptable level. The system of management of food safety needs to be monitored continuously to check that the measures adopted are able to contain the hazard. The aim of monitoring is to assess whether the control chosen to manage a hazard is occurring in practice. Monitoring is a checking and confirming process that ensures a hazard is being managed. The food safety program must identify how each control measure will be monitored; this includes monitoring support programs. For each monitoring action, the food safety program must identify the frequency, type, areas, and indicators to be monitored. This also helps to assess the need for appropriate corrective actions to control the hazard. When a hazard is found to be not under control, corrective action has to be taken. If monitoring finds that the control step in place to manage a hazard is either not working or is not being followed, corrective action must be taken. A corrective action generally consists of two stages, each addressing the issues concerning the product and the process separately. Firstly, immediate action needs to be taken regarding any food that may be unsafe because the hazard is not under control (i.e., addressing the product). Secondly, there needs to be an investigation into the cause of the "loss of control" of the hazard so that steps can be taken to make sure this loss of control does not happen again (i.e., addressing the process). Constant and continuous review of the program by the food business is essential to ensure the safety of the food. The food safety program must include information about the scope of the review. The person who undertakes the review should be someone familiar with the food safety program and the food business' operations, and who has the authority to check records and act on the outcomes. The food business should conduct a review of its food safety program at least once a year. However, in the event that there is any change in the business' food handling activities, or other matters occur that may impact on the food safety program, this review may be required to be undertaken earlier. The scope of the review should describe the food handling operations covered by the review, the procedures and records to be checked, and whether any equipment needs to be checked for accuracy. Maintenance of records as evidence is the most important step in food safety management. Appropriate records are to be kept by the food business, demonstrating action taken to ensure the business is complying with the food safety program. These records must provide sufficient information to show that the business is complying with the food safety program. All records must be kept at least until the next audit of the food safety program. Records can be kept electronically, provided the auditor can access them.

Hazard analysis and control is the key to the success of any food safety management system. Organizations have tried to use certain systems—mainly hygienic and sanitation practices—to establish a food safety program, but these were not comprehensive. Then the concept of identifying hazard points and control measures started gaining importance. Around the year 2000, many countries developed Hazard Analysis and Critical Control Points (HACCP) standards with different levels of certificates, and subsequently the level of implemented HACCP systems based on these standards differed substantially from one country to another. India uses the Indian standard of food hygiene HACCP system and guidelines for its application: IS 15000: 1998. Singapore applies the Singapore Standard 444, while South Africa uses SABS 0330: Code of Practice for the Implementation of a HACCP System. The Netherlands uses Requirements for a HACCP-based Food Safety System (September 2002). Ukraine has the National Standard of Ukraine 4161-2003 Food Safety Management System. Turkey uses the Turkish Standard TS 13001 (March 2003). The Food and Agriculture Organization/World Health Organization (FAO/WHO) Codex Alimentarius has recommended the HACCP Code of Practice CAC/RCP 1-1969, Rev. 4 2003, which has been applied for certification purposes (IFS).

5.2   MANDATORY VERSUS VOLUNTARY APPROACHES TO FOOD SAFETY

Growing concerns about consumer protection and global competitiveness, both of which are closely linked to food quality and safety, have resulted in an ever-expanding number of standards and regulations being released by manifold organizations. Mandatory standards are those which are set by governments in the form of acts or regulations through the process of law. The legislative assembly or parliament of the country approves these regulations. Generally, the term "food law" is used to apply to legislation that regulates the production, trade, and handling of food, and hence covers the regulation of food control, food safety, and relevant aspects of food trade. Minimum quality requirements are included in food law to ensure the foods produced are unadulterated and are not subject to any fraudulent practices intended to deceive the consumer. In case of noncompliance, these are enforced by liability rules and penalties. These are enacted to safeguard the interests of the public and consumers so that they can get safe and wholesome food. Legal processes and penal provisions are available to enforce compliance. Mandatory safety standards are made for products that are likely to be especially hazardous. In making mandatory safety standards, the government protects consumers by specifying the minimum requirements that products must meet before they are supplied. Safety standards require goods to comply with particular performance, composition, contents, methods of manufacture or processing, design, construction, finish, or packaging rules.

In contrast, voluntary standards are set by various stakeholders to harmonize national food safety regulations or to meet specific attributes. These are adopted voluntarily and the objective is to have credibility and win the confidence of the consumers. Voluntary standards are published documents that set out specifications and procedures designed to ensure products, services, and systems are safe, reliable, and consistently perform as intended. Voluntary standards establish a common language that defines the quality and safety criteria.

In line with the globalization of food markets, different levels of standards have to be observed, be they mandatory or voluntary. Four levels of standard ruling/standard setting organizations can be distinguished:

1. *National standard setting organizations*: Almost all countries have acts, laws, or regulations on food safety. If a country does not have its own regulations, it will have adopted the legislations of other countries or will follow the Codex or similar standards. A few of these laws include the U.S. Food Modernization Act, the Indian Food Safety and Standards Act, and the Australia and New Zealand Food Standards Code. These laws are applicable to these countries only. If any food product is to be manufactured, stored, sold, or imported by a country, it must conform to the regulations and laws adopted by that country.

2. *Multilateral standard ruling and multilateral standard setting organizations*: These standards may not have legal bindings, but since they are decided and agreed upon mutually at the international level, they are basically accepted

as norms for trade. An example of a multilateral standard ruling organization is the World Trade Organization (WTO), and an example of a multilateral standard setting organization is the Codex Alimentarius.

3. *Supranational standard setting organizations*: These are voluntary standards adopted and accepted by a few blocs of countries (e.g., trading blocs such as the European Union (EU) and Association of South East Asian Nations (ASEAN)), which are thus accepted as established norms or rules by those countries. These are adopted to have uniformity and basic quality and safety parameters. They have better enforcement and compliance as compared to the Codex but are below the mandatory laws.

4. *Private industry and trade*: These are purely voluntary standards that are adopted by organizations to create and establish the confidence of consumers and address concerns about the quality of the product and that are required for the branding of products. Examples include collective and corporate standards such as the Global Food Safety Initiative (GFSI) and the International Organization for Standardization (ISO).

When a government decides to develop a mandatory standard, a voluntary standard often exists in which experts have already identified ways to address the safety problem. If this has occurred, the government may make all or part of the voluntary standard mandatory. There are two key differences between voluntary and mandatory standards:

1. Mandatory standards are law, and there are penalties and consequences for supplying products that do not comply with these; on the other hand, it is legal to supply products in any country even if they do not meet voluntary standards prevalent in that country.

2. Mandatory safety standards address essential safety features, while voluntary standards may address a range of issues (mainly quality).

Due to globalization of trade, mandatory and voluntary standards are becoming increasingly interlinked. The differences in requirements in all standards are narrowing down and harmonization between standards is increasing. This is one of the reasons that standards set or ruled by multilateral bodies are having an increasing impact on standardization policies at other levels. Standards elaborated by the FAO/WHO Codex Alimentarius Commission (CAC), the International Plant Protection Convention, the World Organisation for Animal Health, and the ISO are recognized by the WTO, which by itself is not a standard setting but a standard ruling organization. Members of the WTO have to adapt their standardization policies at multilateral and national levels based on these references. Although voluntary, standards elaborated by the ISO have become an integral part of an increasing number of standards at all levels. The ISO's work is regulated by the organization's own procedures and also conforms to the WTO's "Code of Good Practice for the Preparation, Adoption and Application of Standards." The same applies to several codes of good practice established by the Organisation for Economic Cooperation and Development, the International Electrotechnical Commission (IEC), and the United Nations Economic Commission for Europe. Some of the voluntary standards, which have wider acceptability among all stakeholders, are increasingly becoming essential requirements for producers, processors, and distributors as their importance for competitiveness in international markets has significantly increased over time. As a consequence of these increasing interdependencies, the distinction between different standard setting levels has become volatile, and the distinction between mandatory and voluntary standards has become irrelevant in practice.

The recent trend of increased government regulation of food safety stands in contrast to the trend in other areas that has greater reliance on voluntary approaches. Consumers also expect more government involvement in food safety as compared to other sectors. There are different reasons or rationales for adopting standards that range from direct damage/harm, market opportunities, to economic considerations. In most pollution control contexts, the likelihood of damage is to damages are to third parties (rather than the consumers of the firm's product) and hence there is less opportunity for market forces (i.e., demand responses) to provide incentives for voluntary adoption of protective measures. Instead, the inducement for voluntary adoption has come primarily from financial incentives (e.g., subsidies) or the threat of imposition of possibly more costly mandatory controls or penal actions. In the context of food safety, however, there is a greater potential for the market to provide adoption incentives when consumers are aware of the safety characteristics of individual products (Segerson, 1998). While the market may work well to induce voluntary adoption of food safety measures for search goods (paper) and experience goods (wine), the above results also suggest that it will not work well for credence goods (vitamin supplements, etc.). In particular, if consumers are unaware of or even simply underestimate potential damages then, even when producers are fully aware, anything less than full liability will lead to overproduction of the good and underprovision of food safety. Since in practice it is unlikely that firms will always be held fully liable even under a strict liability rule (due, for example, to the difficulty of proving causation for credence goods), it is unlikely that firms would invest in an efficient level of protection simply in response to market forces. Thus, in the case of credence goods, adequate consumer protection is likely to be achieved only with some form of government intervention. However, this does not necessarily imply that mandatory regulations must be imposed. Even in the absence of market-driven incentives for investment in food safety, firms might still choose to invest voluntarily if induced to do so by a "carrot" or "stick." Through either government-financed inducements or the threat of possibly more costly mandatory controls, firms can be induced to undertake protective measures voluntarily. If this approach is unsuccessful, however, the government must be prepared to follow through on its threat and impose mandatory standards if adequate food safety is to be ensured.

The food standards can also be classified according to the sectors covered, and these can be placed in three categories:

1. *Global food safety standards for the primary sector.* These address the primary sector or the raw material used for the food industry, which is mainly agriculture and animal husbandry produce. Some of these are the GLOBALG.A.P Integrated Farm Assurance Scheme and Produce Safety Standard, the Canadian Horticultural Council On-Farm Food Safety Program (CanadaGAP), and the Global Aquaculture Alliance (GAA) Seafood Processing Standard.

> Types of food standards

S. No.	Type	The Area issues addressed	Examples
1	For primary sector	Agriculture/aquaculture at farm level/ raw material	GLOBALG.A.P., Canada GAP, GAA
2	For food industry	Manufacturers and processing, storages, transport, etc.	GRMS 4.1, FSSC 22000 IFS version 6 BRC issue 6
3	Single unified standards	Raw material management, etc. (Comprehensive and inclusive)	SQF code 7th edible level 2

FIGURE 5.1   Types of food standards.

2. *Food safety standards for the food industry*: These are standards for the main manufacturing and service sector dealing with the production, handling, storing, and transporting of the food material/product. These include the British Retail Consortium (BRC) Global Standard for Food Safety Issue 6, the BRC/IOP Global Standard for Packaging and Packaging Materials Issue 4, the International Featured Standard (IFS) Food Version 6, the Global Red Meat Standard (GRMS) Version 4.1, and the Food Safety System Certification (FSSC) 22000.

3. *Food safety standards for all sectors in the food industry*: These are comprehensive and inclusive standards. These deal with all the raw materials and sectors dealing with food and have been prepared to assert that an organization should have a single standard in place of two or three. These include the Safe Quality Food (SQF) Code 7th Edition Level 2.

Figure 5.1 illustrates various types of food standards prevalent in the industry. The list may not be inclusive and exhaustive as there are variations, repetitions, and segregations in each type, as well as scope for improvement and innovations.

5.3   NATIONAL FOOD SAFETY REGULATIONS, LAWS, AND STANDARDS

Food safety legislation should address the complete food supply chain so that food ready for consumption is safe. The establishment and updating of food safety legislation throughout the food chain is essential for establishing an effective food safety system. Food safety legislation should be developed and updated taking into consideration the following:

1. Specific needs of consumers and food producers
2. Developments in technology
3. Emerging hazards
4. Changing consumer demands
5. Social, religious, and cultural habits
6. New requirements for trade
7. Harmonization with international and regional standards
8. Obligations under WTO agreements

In countries around the world, the overarching principles concerning food safety and consumer protection are established in national legislation. Yet the current policy environments in many countries do not always give it significant impetus. In some of the countries, formulation and implementation of food safety laws and regulations are often fragmented. Sometimes these are enforced by different agencies and also cover areas and sectors in isolation and separately. Due to this lack of a holistic and coordinated approach to food safety, the enforcement of regulations also becomes difficult. Due to the confusion prevailing as a result of such situations, even the subject matter often misses the attention of policy makers and pressure groups. Food safety makes headlines only when mishaps go beyond control and consume valuable human lives.

5.3.1 THE UNITED STATES AND FOOD SAFETY LEGISLATION

The United States was at the forefront of the food safety movement, along with the countries of the EU. It was due to consumer pressure groups changing the aspirations of consumers, progress in food preservation, emergencies related to food-related health issues, and technology developments that legislations were enacted and improved from time to time. The Food and Drug Administration (FDA) Food Safety Modernization Act, passed by Congress on December 21, 2010, aims to ensure that the U.S. food supply is safe by shifting the focus of federal regulators from responding to contamination to preventing it.

The FDA has comprehensive, prevention-based controls across the food supply:

- The legislation transformed the FDA's approach to food safety from a system that far too often responds to outbreaks to one that prevents them. It does so by requiring food facilities to evaluate the hazards in their operations, implement and monitor effective measures to prevent contamination, and have a plan in place to take any corrective actions that are necessary.
- The legislation also requires the FDA to establish science-based standards for the safe production and harvesting of fruits and vegetables to minimize the risk of serious illnesses or death.
- This new ability to hold food companies accountable for preventing contamination is a significant milestone in the efforts to modernize the food safety system.

- The legislation recognizes that inspection is an important means of holding the industry accountable for its responsibility to produce safe products.
- The legislation provides significant enhancements to the FDA's ability to achieve greater oversight of the millions of food products coming into the United States from other countries each year. An estimated 15% of the U.S. food supply is imported, including 60% of fresh fruits and vegetables and 80% of seafood.
- For the first time, the FDA will have mandatory recall authority for all food products.
- The legislation recognizes the importance of strengthening existing collaboration among all food safety agencies—federal, state, local, territorial, tribal, and foreign—to achieve public health goals.
- It also recognizes the importance of building the capacity of state, local, territorial, and tribal food safety programs.

5.3.2 FOOD SAFETY REGULATIONS IN THE EUROPEAN UNION

All member countries of the EU are following dual legislations in food safety, especially because of their situation (EU being cooperation of sovereign countries) and facilitating the free flow of food trade. In all member states of the EU, principles concerning food safety and consumer protection are established in national legislation. The Food Standards Agency of the United Kingdom represents England, Wales, and Northern Ireland on food safety and standards issues in the EU. The agency consults widely in seeking to help develop a framework of well-founded, proportionate, and effective European food law, taking into account the principles of better regulation. As well as working in close collaboration with other government departments, the agency has regular contact with the Council of the EU, the Commission, Standing Committees, the European Parliament, and the European Food Safety Authority (EFSA). The EFSA is an independent source of scientific advice on existing and emerging risks associated with the food chain. In Europe, risk assessment is separate from risk management. As the risk assessor, the EFSA produces scientific opinions and advice to provide a foundation for European policies and legislation and to support the European Commission, European Parliament, and EU member states. The EFSA works in close collaboration with national authorities and in open consultation with its stakeholders. In Germany, the Lebensmittel-, Bedarfsgegenstände- und Futtermittelgesetzbuch (LFGB) comprises the basic German food and feed law, most of which is based on fully harmonized EU regulations and directives. The LFGB provides the basic definitions, procedural rules, and goals of German food law. It defines general food safety and health protection rules, addresses labeling requirements, and regulates inspection, detention, and seizure rules of suspect food. These rules apply to domestically produced and imported food products alike. Many standards in food hygiene have been harmonized in the Single European Market. Detailed information on the applicable legal framework is available on the website of the European Directorate General for Health and Consumers. Nevertheless, several stricter food safety standards apply on the national level. Due to the free movement of goods, even products that do not comply with these national requirements may enter the country, as long as they are legal in another EU member state. However, at EU level, food legislation has evolved, with some of these basic principles having been established in an overarching legal instrument. Regulation (EC) No 178/2002 of the European Parliament and of the Council of 28/1/2002[5] lays down the general principles and requirements of food law, establishing the EFSA and laying down procedures in matters of food safety:

- This regulation ensures the quality of foodstuffs intended for human consumption and animal feed.
- It guarantees the free circulation of safe and secure food and feed in the internal market.
- EU food legislation protects consumers against fraudulent or deceptive commercial practices.
- The legislation aims to protect the health and well-being of animals, plant health, and the environment.

The objective of the EU's food safety policy is to protect consumer health and interests while guaranteeing the smooth operation of the single market. In order to achieve this objective, the EU ensures that control standards are established and adhered to as regards food and food product hygiene, animal health and welfare, and plant health and preventing the risk of contamination from external substances. It also lays down rules on the appropriate labeling of these foodstuffs and food products. This policy underwent reform in the early 2000s, in line with the "From the Farm to the Fork" approach, thereby guaranteeing a high level of safety for foodstuffs and food products marketed within the EU at all stages of the production and distribution chains. This approach involves both food products produced within the EU and those imported from third-party countries. It also guarantees the free circulation of safe and secure food and feed in the internal market. In addition, the EU's food legislation protects consumers against fraudulent or deceptive commercial practices. This legislation also aims to protect the health and well-being of animals, plant health, and the environment. In the EU, the application of a HACCP approach in the food industry is compulsory by law. This also applies to countries exporting to the EU, according to Regulation (EC) 852/2004 on the hygiene of foodstuffs.

5.3.3 FOOD STANDARDS AUSTRALIA NEW ZEALAND

Food Standards Australia New Zealand (FSANZ) is an independent statutory agency established by the Food Standards Australia New Zealand Act 1991. FSANZ develops standards that regulate the use of ingredients, processing aids, colorings, additives, vitamins, and minerals. The act also covers the composition of some foods (e.g., dairy, meat, and beverages), as well as standards developed by new technologies such as genetically modified foods. FSANZ is responsible for some labeling requirements for packaged and unpackaged food (e.g., specific mandatory warnings or advisory labels). Overarching food policy is set by the ministers in Australia and New Zealand who are responsible for food regulation. These ministers make up the Australia and New Zealand Ministerial Forum on Food Regulation, which develops food regulatory policy and policy guidelines that FSANZ must have regard for when setting food standards. FSANZ develops the food standards in the Food Standards Code with advice from other government agencies and input from stakeholders. Food standards are enforced by the states and territories (usually their health or human services departments) or, in some cases, by local government. These authorities regularly check food products for compliance with the Food Standards Code. Australian state and territory and New Zealand government agencies are responsible for implementing, monitoring, and enforcing food regulation through their individual food acts and other food-related legislation. These agencies vary between jurisdictions. The Department of Agriculture enforces the Food Standards Code at the border in relation to imported food. Food regulation authorities in Australia and New Zealand work together to ensure food regulations are implemented and enforced consistently. This work is done through the Implementation Subcommittee for Food Regulation, through face-to-face meetings, out-of-session business, and separate collaborations.

5.3.4 Codex Alimentarius

A collection of internationally recognized standards, codes of practice, guidelines, and other recommendations relating to foods, food production, and food safety are collected in the Codex Alimentarius (Latin for "book of food"). The Codex texts are developed and maintained by the Codex Alimentarius Commission (CAC), a body that was established by the FAO and the WHO in 1962 and that held its first session in Rome in October 1963. The Commission's main goals are to protect the health of consumers and ensure fair practices in the international food trade. The Codex is recognized by the WTO as an international reference point for the resolution of disputes concerning food safety and consumer protection. The Codex covers all foods, whether processed, semiprocessed, or raw. Several standards for specific foods have been developed:

- Meat products (fresh, frozen, processed meats, poultry)
- Fish and fishery products (marine, fresh water, aquaculture)
- Milk and milk products
- Foods for special dietary uses (including infant formula and baby foods)
- Fresh and processed vegetables, fruits, and fruit juices
- Cereals and derived products, dried legumes
- Fats, oils, and derived products such as margarine
- Miscellaneous food products (chocolate, sugar, honey, mineral water)

In addition to standards for specific foods, the Codex contains general standards covering matters such as food labeling, food hygiene, food additives, contaminants in foods, and pesticide and veterinary chemical residues, as well as procedures for assessing the safety of foods derived from modern biotechnology. It also contains guidelines for the management of official (i.e., governmental) import and export inspection and certification systems for foods. It is a voluntary reference standard for food and there is no obligation for countries to adopt Codex standards as a member of either the CAC or any other international trade organization. However, the use of the Codex during international disputes does not exclude the use of other references or scientific studies as evidence of food safety and consumer protection.

5.3.5 INTERNATIONAL FOOD SAFETY AUTHORITIES NETWORK

The International Food Safety Authorities Network (INFOSAN) is a global network of national food safety authorities managed jointly by two United Nations agencies: the FAO and the WHO. Through INFOSAN, the WHO assists member states in managing food safety risks, ensuring rapid sharing of information during food safety emergencies to stop the spread of contaminated food from one country to another. Such collaboration is necessary as the increased globalization of the food trade has increased the risk of contaminated food spreading quickly around the globe. INFOSAN also facilitates the sharing experiences and tested solutions in and between countries in order to optimize future interventions to protect the health of consumers. The national authorities of 181 member states are part of the network.

5.3.6 FOOD SAFETY AND ASIA

Food safety is a key issue for consumers in Asia. With fast-growing economies, a burgeoning middle class, and complex supply chains, countries in Asia face a growing array of food safety challenges. These are giving rise to innovative solutions and collaborative initiatives between governments and the private sector across the entire region. The high cost of implementation and poor enforcement are the main reasons for poor food safety controls. There is great variation between Asian countries due to disparities in development and the awareness of consumers. While Japan, Singapore, and South Korea have better compliance records, a few countries do not have even their own food safety regulations and have adopted these from other countries. India and China are two of the main countries that have very large populations to feed and are also the largest producers of many of the foods in the world. The Chinese government attempted to consolidate food safety regulation with the creation of the State Food and Drug Administration of China in 2003. In March 2013, the national regulatory body was rebranded and restructured as the China Food and Drug Administration, elevating it to a ministerial-level agency (Prakash, 2013).

In India, the Food Safety and Standards Authority of India (FSSAI) acts as a single reference point for all matters relating to food safety and standards in the country. The FSSAI has been mandated by the Food Safety and Standards Act 2006 for framing regulations to: lay down standards for articles of food; lay down guidelines for the accreditation of bodies engaged in the certification of food businesses; provide technical support to central and state governments in framing policies for food safety and nutrition; manage data regarding food consumption, incidence, and prevalence of biological risk, contaminants in food, and residues of various contaminants in food products; identify emerging risks and introduce a rapid alert system; create an information network across the country so that the public receives information about food safety and issues of concern; and promote general awareness about food safety and food standards.

5.3.7   ASEAN Food Safety Network

The ASEAN Food Safety Network (AFSN) was established according to the resolution of the 25th meeting of the ASEAN Ministers on Agriculture and Forestry (Malaysia, 2003). The purpose of the establishment of the AFSN was for it to be a channel for ASEAN member states to exchange information relevant to food safety. Consequently, Thailand developed the website, which was launched at www.aseanfoodsafetynetwork. net in September 2004. So far Thailand, as a coordinator of the AFSN, has worked closely with good effort and cooperation of all national focal points assigned by individual member states. At present, the AFSN serves as a central platform for coordination and information exchange for ASEAN cooperation on food safety under ASEAN Ministers, namely the ASEAN Ministers on Agriculture and Forestry, the ASEAN Economic Ministers, and the ASEAN Health Ministers. The current works of the AFSN include systems and activities for member states to exchange information as well as to work and communicate electronically via the network. These channels include the ASEAN Consultative Network, as well as websites for a number of ASEAN working groups, including the Expert Group on Food Safety, the Task Force on Codex, and the Working Group on Halal Foods. Furthermore, the AFSN has offered and disseminated food safety regulations and requirements, relevant information, or situations that have been recently issued, implemented, or notified by member states, other countries, and international organizations. Therefore, member states are able to access and obtain recent information in order to prepare themselves for newly implemented requirements or urgent occurrences. This assists the ASEAN member states in maintaining and enhancing their potential and competitiveness in a global market (FIA, 2012).

5.3.8   Africa and Food Safety

There are many challenges to food safety in Africa, but the continent has to establish safety standards before it can attain the United Nations MDGs of eradicating extreme poverty and hunger and combating HIV/AIDS, malaria, and other diseases. The continent of Africa consists of 54 very diverse countries. Covering 11.7 million square miles—the landmass of the world's second largest and second most populous continent—Africa has a number of countries that rely predominantly on subsistence farming and street food vending to feed their populations. Other African countries are more developed to varying degrees. Malfunctioning regional trade is one of the major reasons why so many African countries are net importers of food and feed, and why the region is vulnerable to volatile global prices. At the same time, existing opportunities for agro-ecological complementarities within countries and between subregions remain untapped, despite possibilities to exploit them for a better balance between food-surplus and food-deficit areas. The key role that cross-border trade and regional cooperation plays in the development of national economies and the competitiveness of economic sectors is increasingly recognized. The formalization of the African Union in 2002 was a fresh impulse for the further development of the long time existing Regional Economic Communities (RECs) in Sub-Saharan Africa. In particular, the East

African Community (EAC) and the Common Market for Eastern and Southern Africa (COMESA) can report successes with advanced economic integration. Representing a crucial factor of value chain competitiveness, compliance with regulations and standards for food safety and quality is a prerequisite for accessing markets. This is not only true for opening and maintaining shares in global markets but equally and increasingly for assuring consumer protection while competing with imports in domestic markets, and for seizing still largely untapped opportunities in regional markets. Just like the urgent need to improve public and private sector capacities for the surveillance of and compliance with national regulations and standards, regional harmonization is decisive for the competitiveness in intra- and interregional trade within and among the RECs. The EAC and COMESA have both initiated the development of common sanitary and phytosanitary protocols. Aiming at promoting intra- and interregional trade, both RECs have also started to develop the necessary quality infrastructure and to build the human capacities for compliance with international standard frameworks. Furthermore, with the East African Standards compendium, the EAC has developed a comprehensive set of harmonized standards, which COMESA has decided to largely adopt in order not to duplicate. There is an urgent need to develop, revise, and harmonize standards/regulations and strengthen the capacities of the national and regional quality infrastructure, and also to address organizational and capacity upgrading needs at the food supply chain level. Without developing the capacities of value chain operators, all efforts will fail to increase regional trade and to seize opportunities from existing regional competitive advantages for the sake of food security and pro-poor growth.

5.3.9 FOOD SAFETY IN SOUTH AMERICA

South America is strong in natural resources for food production. Covering an area of 6.89 million square miles, South America is home to an estimated 387.5 million people. The continent is also strong in innovative approaches on the part of large food exporters, as they typically respond positively and efficiently to trends in global food markets. Food is produced in different seasons there compared to northern-hemisphere continents, and this contributes favorably to the consistency of foreign food supplies year round. South American countries export a wide variety of foods to the United States, including fresh fruit, salmon, beef, wine, and coffee. This ensures that these countries have to follow strict food safety requirements so as to meet the standards of U.S. market and also other countries. Market access all over the world and brand protection have been the major drivers of food safety in South America. As a consequence of globalization, the actualization and implementation of new food legislation, as well as the modernization of food safety inspection agencies to comply with international requirements and guidelines, have been on the agenda of most countries in the region in recent years. Probably due to the importance of food safety for growing food exports, this has defined a very close relationship between food safety and agricultural development.

The Union of South American States (UNASUR), a regional organization composed of 12 South American countries, has been working on reducing inequality and promoting health since its foundation in 2008. Put together by some of the countries suffering from the highest levels of economic inequality in the world, the focus is partly on trying to tackle such disparity. Furthermore, because this body was founded by a group of left-leaning governments, it approaches health as a social issue that should be upheld to the same value as a human right. At the level of policy implementation, UNASUR is coordinating new transnational risk-mitigation projects and funding for health and food security programs in order to address the health risks associated with unhealthy foods, chronic noncommunicable diseases, and sedentary lifestyles. Measures to be taken at a regional level include assuring safety throughout the entire food chain by enforcing science-based regulations and applying risk analysis focusing on those critical points in the food chain that require closer monitoring by the relevant government authorities, and improving coordination both among countries in the region and between food regulators and food producers in each country. Still, not all countries in the region have harmonized their food legislation with Codex standards. This makes exporting food to the rest of the world more difficult. Food legislation is currently enforced by a number of ministries, including, depending on the country, those for agriculture, health, economy, tourism, trade, and industry, among others. With more than one implementing agency, there is often the risk that the relevant regulations overlap. In addition, these regulations are often outdated and are not science based. Moreover, food safety and quality reference laboratories do not exist in all countries of the region. The 200 experts underlined the necessity of ensuring that all countries in the region have their own food testing facilities or have access to suitable ones in the region.

5.4   PREREQUISITE PROGRAMS

The WHO defines a prerequisite program as the "practices and conditions needed prior to and during the implementation of HACCP and which are essential for food safety." Prerequisite programs (PRPs) provide a foundation for an effective HACCP system. These are often facility-wide programs rather than process or product specific. They reduce the likelihood of certain hazards being present in food. PRPs are the universal steps or procedures that control the operational conditions within a food establishment. They are designed to create an environment favorable to the production of safe food. They are the basic conditions and activities that are necessary to maintain a hygienic environment. The following areas must be taken into consideration when implementing PRPs:

- Construction and layout of the building
- Layout of premises, workspaces, and employee facilities
- Water management
- Waste management
- Equipment (cleaning, maintenance/preventive maintenance)
- Management of purchased materials
- Measures to prevent cross-contamination
- Cleaning and sanitizing
- Pest control
- Personal hygiene
- Product recall and traceability
- Transportation and storage

Hygiene standards and procedures, usually described as "good hygienic practices" or "good manufacturing practices," have been in place for many years and constitute an essential tool in traditional food control. These concepts are still essential in a modern food control system for providing the basic environmental and operating conditions for the production of safe food, and thus they are a requisite or foundation for HACCP in an overall food safety management program. A proper and well-designed PRP allows the food safety management team to focus and concentrate on the hazards directly applicable to the product and the processing procedures without undue consideration and repetition of protection from hazards from the surrounding environment. It is important to point out that the PRP certainly relates to safety and therefore is an essential part of the total quality assurance program. Thus, part of the PRP (e.g., sanitation control) must lend itself to all aspects of a critical control point (CCP), such as establishing critical limits, monitoring, taking corrective actions, recordkeeping, and creating verification procedures. However, occasional deviation from a PRP requirement would not by itself be expected to create a food safety hazard of concern. Therefore, deviations from compliance in a PRP usually do not result in a reaction against the product. This is in contrast to a CCP, where any deviation from the established critical limits always leads to reaction against the product. PRPs have become an essential component of any good food safety standard. All major standards will have PRPs, and some of these are as follows:

1. For FSSC 22000 the requirements for PRPs can be found in (FSSC 22000)
   a. ISO/TS 22002-1, for food manufacturing
   b. PAS 223, for food packaging manufacturers
   c. PAS 222, for feed producers
2. For ISO 22000 the requirements for PRPs can be found in
   a. ISO 22000, section 7.2
   b. ISO/TS 22002-1, which provides more detail for PRPs that may be used to set up programs
3. For BRC the requirements for PRPs can be found in (BRC)
   a. BRC global standard for food safety, Sections 4–7, for food manufacturing
   b. BRC global standard for food packaging, for food packaging manufacturing
4. For SQF Edition 7 the requirements for PRPs can be found in Modules 3–15, for example
   a. Module 11, for food manufacturing
   b. Module 13, for food packaging manufacturers

Although there are differences in the specific requirements of each of the standards mentioned, all these programs address key elements for maintaining hygienic conditions in the manufacturing environment. Food safety management systems (FSMS) are intended to provide organizations with the elements of an effective food safety system in order to achieve the best practices in food safety and to maintain economic goals. The intention is that these systems may be integrated with other management requirements such as those for quality and environment. In response to customer demands for a recognized standard against which a FSMS can be audited and certified, the ISO 22000:2005 standard ("Food safety management systems—Requirements for any organization in the food chain") has been developed, which represents a major step forward in the harmonization of the requirements for food safety management on a global level.

5.5   GLOBAL FOOD SAFETY INITIATIVE

Some countries have become more demanding regarding the safety of food and the requirements for the importing of food products from developing countries. The national food safety standards of various developing countries and the respective primary producers and food processing companies that have been certified against these national standards are not considered adequate for international trade as requirements have been made more stringent. Therefore, the private sector has taken action by introducing modern food safety systems such as the benchmarked Food Safety Standards for the Primary Sector and the Food Industry. In this process, retail organizations play an increasingly important role in demanding that their suppliers conform to internationally recognized food safety standards. Against this background, the Global Food Safety Initiative (GFSI) was launched in 2000, following a number of food safety crises when consumer confidence was at an all-time low. Its collaborative approach to food safety brings together international food safety experts from the entire food supply chain at technical working group and stakeholder meetings, conferences, and regional events to share knowledge and promote a harmonized approach to managing food safety across the food industry. The GFSI was founded to deliver equivalence and convergence between effective food safety management systems through its benchmarking process, and continues to flourish in doing so. Benchmarking is a "procedure by which a food safety-related scheme is compared to the GFSI Guidance Document." The benchmarking process determines equivalency against an internationally recognized set of food safety requirements, based on industry best practice and sound science. These requirements are developed through a consensus-building process by key stakeholders in the food supply chain and can be found in the GFSI Guidance Document Sixth Edition, freely available for download (GFSI, 2011).

The GFSI is managed by the Consumer Goods Forum, a global, parity-based industry network driven by its members. On January 5, 2011, the GFSI released the sixth edition of its Guidance Document, which contains requirements for food safety schemes for the primary sector and the food industry. Scheme owners who had previously been benchmarked against the fifth edition, as well as new scheme owners, were invited to submit a benchmarking application and apply for recognition against the GFSI Guidance Document Sixth Edition (GFSI, 2011).

The GFSI benchmarking/recognized schemes have scopes as follows:

AI	Farming of Animal Products	FI	Feed Production—Single Feed
AII	Farming of Fish Products	FII	Feed Production—Compound Feed
BI	Farming of Plant Products	G	Catering
BII	Farming of Grains and Pulses	H	Retail/Wholesale
C	Pre-Processing of Animal Products	I	Provision of Food Safety Services
D	Pre-Processing of Plant Products	JI	Provision of Transport and Distribution Services—Perishables
EI	Processing of Animal Perishable Products	JII	Provision of Transport and Distribution Services—Ambient
EII	Processing of Plant Perishable Products	K	Processing Equipment Manufacture
EIII	Processing of Animal and Plant Perishable Products (Mixed Products)	L	Production of (Bio)Chemicals
EIV	Processing of Ambient Stable Products	M	Production of Food Packaging
		N	Food Broker/Agent

5.6   STANDARDS FOR THE PRIMARY SECTOR

5.6.1   GLOBALG.A.P.

The GLOBALG.A.P. flagship standard is the Integrated Farm Assurance (IFA) standard that covers crops, livestock, and aquaculture and emphasizes a progressive, holistic approach to farm certification. It was started in 1997 as EUREPG.A.P., an initiative by retailers belonging to the Euro-Retailer Produce Working Group, and it was renamed GLOBALG.A.P. in 2007. The objective of EUREPG.A.P. was to raise standards for the production of fresh fruit and vegetables. The prepared document (checklist) sets out a framework for good agricultural practice (GAP) on farms. It defines essential elements for the development of best practice for the global production of horticultural products (e.g., fruits, vegetables, potatoes, salads, cut flowers, and nursery stock, as well as tea). The GLOBALG.A.P. is an affiliate organization of a not for-profit trade association with the crucial objective of providing safe, sustainable agricultural production worldwide. It sets voluntary standards for the certification of agricultural products around the globe, and more and more producers, suppliers, and buyers (currently over 100 members on every continent) are harmonizing their certification standards to match it. The EUREPG.A.P. standards helped producers comply with Europe-wide accepted criteria for food safety, sustainable production methods, worker and animal welfare, and responsible use of water, compound feed and, plant propagation materials. Over the next 10 years the process spread throughout the continent and beyond. Driven by the impacts of globalization, a growing number of producers and retailers around the globe joined in, gaining the European organization global significance. To reflect both its global reach and its goal of becoming the leading international GAP standard, EUREPG.A.P. changed its name to GLOBALG.A.P. The certification criteria of the fruit and vegetables scope of the IFA standard focus on:

- Food safety and traceability
- Environmental sustainability
- Worker operational health and safety
- Integrated pest management, quality management system, and basic HACCP

The Produce Safety Standard only focuses on the food safety and traceability elements of the IFA standard. This standard is a subset of the IFA standard that was developed for the North American market, where the demand for compliance with the food safety elements of the IFA standard has priority over the non-food safety components. The GLOBALG.A.P. IFA Scheme (Sub-Scope Fruit and Vegetables) Version 4 and the Produce Safety Standard Version 4 (Scope Extension) have been successfully benchmarked by the GFSI and recognized against the following scopes of the GFSI Guidance Document Sixth Edition (April 24, 2013):

BI	Farming of Plants
D	Pre-Process Handling of Plant Products

5.6.2 CANADIAN HORTICULTURAL COUNCIL ON-FARM FOOD SAFETY PROGRAM

The Canadian Horticultural Council On-Farm Food Safety Program (CanadaGAP) is a food safety certification program for companies that produce, pack, and store fresh fruits and vegetables. Launched in 2008 by the Canadian Horticultural Council, the scheme has been owned and operated by not-for-profit Canadian corporation CanAgPlus since November 2012. The standard comprises two manuals—one specific to greenhouses, the second for other fruit and vegetable operations—developed in consultation with the horticultural sector and reviewed annually for technical soundness by Canadian government officials. The manuals are based on a rigorous hazard analysis, applying the seven principles of the internationally recognized HACCP approach. In 2010, Options B and C of the CanadaGAP scheme were benchmarked to Version 5 of the GFSI Guidance Document. The program now has over 2000 participating producers across Canada. The CanadaGAP Scheme Version 6 Options B and C and Program Management Manual Version 3 have been successfully rebenchmarked by the GFSI.

5.6.3 GLOBAL AQUACULTURE ALLIANCE SEAFOOD PROCESSING STANDARD

The GAA is an international, non-profit trade association, organized in 1997 and based in the United States, dedicated to advancing environmentally and socially responsible aquaculture. Its mission is to further develop environmentally responsible aquaculture to meet world food needs. Through the development of its Best Aquaculture Practices Certification Standards, the GAA has become the leading standard setting organization for aquaculture seafood (GAA).

The GAA encourages the use of responsible production systems that are sustainable regarding environmental and community needs, and that efficiently provide safe, wholesome aquaculture products to the world's population. The GAA articulates the importance of aquaculture as a source of food and employment. It supports technological research and provides this information openly to membership and research facilities. The alliance also advocates for the industry regionally and globally, and promotes effective, coordinated government regulatory and international trade policies. The GAA Seafood Processing Standard has been successfully benchmarked by the GFSI and has achieved recognition against the following scopes of the GFSI Guidance Document Sixth Edition (May 16, 2013):

EI	Processing of Animal Perishable Products

5.7 GLOBAL FOOD SAFETY STANDARDS FOR THE FOOD INDUSTRY

5.7.1 BRITISH RETAIL CONSORTIUM GLOBAL STANDARD FOR FOOD SAFETY ISSUE 6

This standard covers food safety and management of product quality in food packing and processing operations. The BRC Global Standard for Food Safety is a checklist for those companies supplying retailer-branded food products. The standard has been developed to assist retailers in their fulfillment of legal obligations and protection of the consumer, by providing a common basis for the inspection of companies supplying retailer-branded food products (BRC Global Standards, 2011).

The BRC Global Standard for Food Safety was one of the original GFSI-benchmarked schemes and is used around the world, with certificates in over 100 countries and in excess of 15,000 certified sites. The standard is owned by the BRC and written and managed with the input of an international multistakeholder group made up of food manufacturers, retailers, and food service and certification body representatives.

The BRC Global Standard for Food Safety Issue 6 has been successfully benchmarked by the GFSI and has achieved recognition against the following scopes of the GFSI Guidance Document Sixth Edition (September 20, 2012):

D	Pre-Process Handling of Plant Products
EI	Processing of Animal Perishable Products
EII	Processing of Plant Perishable Products
EIII	Processing of Animal and Plant Perishable Products (Mixed Products)
EIV	Processing of Ambient Stable Products
L	Production of (Bio)Chemicals

5.7.2 BRC/IOP Global Standard for Packaging and Packaging Materials Issue 4

This is an established standard for the manufacture and conversion of packaging materials for both food and non-food use, with approximately 2000 certified sites around the world. The standard covers the hygienic production of packaging materials and the management of the quality and functional properties of the packaging to provide assurance to customers. The standard is operated by the BRC in conjunction with the Packaging Society and an advisory committee of stakeholders. The BRC Global Standard for Packaging and Packaging Materials Issue 4 has been successfully benchmarked by the GFSI and has achieved recognition against the following scopes of the GFSI Guidance Document Sixth Edition (September 20, 2012) (BRC Global Standards):

M	Production of Food Packaging

5.7.3 International Featured Standard Food Version 6

The IFS Food is a standard for auditing food safety and the quality of the processes and products of food manufacturers. The IFS standard is based on the philosophy of the ISO 9001:2000 standard. The IFS standard (also a checklist), like the BRC standard, involves primarily the set up of the HACCP system. The standard has been in existence since 2003 and is currently operating its sixth version. Over 11,000 certificates in 90 different countries were issued in 2011.

The IFS Food Version 6 has been successfully rebenchmarked by the GFSI and has achieved recognition against the following scopes of the GFSI Guidance Document Sixth Edition (January 4, 2013):

C	Animal Conversion
D	Pre-Process Handling of Plant Products
EI	Processing of Animal Perishable Products
EII	Processing of Plant Perishable Products
EIII	Processing of Animal and Plant Perishable Products (Mixed Products)
EIV	Processing of Ambient Stable Products
L	Production of (Bio)Chemicals

The IFS was created by the Federation of German Distributors (after which it was supplemented by French distributors) in order to initiate a systematic and uniform evaluation of food product suppliers. IFS management has five regional offices worldwide, coordinates technical working groups in different languages (German, French, North American, Spanish, and Italian) with different stakeholders (retailers, industry, certification bodies, and food services), and relies on a continuous improvement process of IFS standards, databases, and integrity programs, among others.

5.7.4 Global Red Meat Standard Version 4.1

The GRMS is a standard specifically developed for the processes of slaughtering, cutting, and deboning, and for the sale of red meat and meat products (e.g., beef and pork). In contrast to other more generic food industry quality schemes, the

GRMS has been tailored to the specific requirements that apply to the meat industry. The standard comprises the entire production chain and is, therefore, applicable to all aspects of transport, lairage, stunning, slaughtering, deboning, cutting, and handling of meat and meat products. The GRMS sets out the requirements for all processes relating to the production of meat and meat products, and focuses on areas critical to achieving the highest safety and quality levels. The GRMS is EN45011 accredited and independently audited. The GRMS was launched in 2006 and first achieved recognition by the GFSI in 2009. The standard has been extensively revised with advice and input from relevant stakeholders, ensuring that it continues to meet the requirements of manufacturers and retailers. Some misunderstandings regarding specific clauses in the Section II: Audit Protocol and Section IV: Requirements for Auditor Qualification, Training and Experience have been clarified.

GRMS Version 4.1 has been successfully rebenchmarked by the GFSI and has achieved recognition against the following scopes of the GFSI Guidance Document Sixth Edition (February 7, 2013):

C	Animal Conversion
EI	Processing of Animal Perishable Products
EIII	Processing of Animal and Plant Perishable Products (Mixed Products)

5.7.5   FOOD SAFETY SYSTEM CERTIFICATION 22000

The Food Safety System Certification (FSSC) 22000 is a robust, ISO-based, internationally accepted certification scheme for the assessment and certification of food safety management systems in the whole supply chain. FSSC 22000 uses the existing standards ISO 22000:2005, ISO 22003, and technical specifications for sector PRPs (ISO). The certification is accredited under ISO Guide 17021 and recognized by the GFSI (Foundation for Food Safety Certification). The non-profit Foundation for Food Safety Certification retains the ownership, copyright, and license agreements for certification bodies. The FSSC 22000 standard combines key food safety elements, prerequisites, and HACCP with the management system (ISO 9000 approach). This includes interactive communication along the food chain and the continual improvement of the food safety management system. The standard will further contribute to the standardization and harmonization of certified HACCP systems worldwide. The FSSC 22000 has been successfully rebenchmarked by the GFSI and has achieved recognition against the following scopes of the GFSI Guidance Document Sixth Edition (February 22, 2013) (FSSC22000, 2013):

C	Animal Conversion
D	Pre-Process Handling of Plant Products
EI	Processing of Animal Perishable Products
EII	Processing of Plant Perishable Products
EIII	Processing of Animal and Plant Perishable Products (Mixed Products)
EIV	Processing of Ambient Stable Products
L	Production of (Bio)Chemicals
M	Production of Food Packaging

5.8 FOOD SAFETY STANDARD FOR ALL SECTORS

5.8.1 SAFE QUALITY FOOD CODE SEVENTH EDITION LEVEL 2

The SQF Code has been redesigned for use by all sectors of the food industry, from primary production to transport and distribution. The Seventh Edition applies to all industry sectors and replaces the SQF 2000 Code Sixth Edition and the SQF 1000 Code Fifth Edition.

The SQF Code was originally designed in Australia in 1999. The code is applied in New Zealand, Australia, North America, South Africa, and the Middle East. This code is less frequently applied in Europe (Safe Quality Food Institute).

The SQF Code is a process and product certification standard. It is a HACCP-based food safety and quality management system that utilizes the National Advisory Committee on Microbiological Criteria for Food and the CAC's HACCP principles and guidelines. The SQF Code is intended to support industry- or company-branded products and offers benefits to suppliers and their customers. With the consistent application of the SQF program by certification bodies that have been accredited to ISO/IEC Guide 65: 1996, products produced and manufactured under SQF Code certification retain a high degree of acceptance in global markets. The SQF Code Seventh Edition Level 2 has been successfully rebenchmarked by the GFSI and has achieved recognition against the following scopes of the GFSI Guidance Document Sixth Edition (October 15, 2013):

A	Farming of Animal Products
BI	Farming of Plant Products
C	Animal Conversion
D	Pre-Process Handling of Plant Products
EI	Processing of Animal Perishable Products
EII	Processing of Plant Perishable Products
EIII	Processing of Animal and Plant Perishable Products (Mixed Products)
EIV	Processing of Ambient Stable Products
L	Production of (Bio) Chemicals
M	Production of Food Packaging

APPENDIX 5.A

CERTIFICATION AND ACCREDITATION FOR A FSMS

The process of certification is adopted to certify conformation with a performance standard that has been adopted and selected by a unit organization. There are a set of rules around the procedure and management for carrying out a conformity assessment. Such an assessment leads to the issuance of a certificate.

Any FSMS will be specific to the type of product(s), location(s), activities, processes, and production sites. The prerequisites to a FSMS are two essentials: one is the food safety policy, and the other is the adoption of the same food safety standard or parameter, which will always be higher than the mandatory standards. After the establishment of a FSMS, audits are conducted.

The major benefit or use of a management-system-based approach to food safety is that it enhances consumer confidence and also improves food quality. This is the reason that individual consumers, consumer groups, and governments are demanding certified food. There is increased use of second-party audit and third-party certification. The genesis of the system of food safety certification was a number of food safety crises that lowered the confidence of consumers.

CERTIFICATION PROCESS

Before any certification process is started, there are preparatory conditions that need to be fulfilled. If the preparation phase is handled well, it becomes easier to implement the FSMS and achieve a certificate. The steps in the preparation phase are as follows:

1. *Identification of certification requirements*: This is the first and most important step, which is crucial for the success of the program. It is required to be aware of the following:
   a. Identify the product categories for which certification is required. Different products are categorized for certification on the basis of similarity in type and food safety. Normally, standards are also set down category-wise so that comparisons can also be made.
   b. Identify the FSMS or standards against which certification is required. This will be decided as per the company requirements for the quality and safety of the food; hence the food safety policy of the company plays an important role. Based on this, the certification body will also be decided. Here the company needs to examine the scope of accreditation of the certification body and check if the auditors are qualified and trained. The sanction cost of an audit also becomes a major factor in deciding upon a certification body.
   c. Identify a team for implementation. The key individuals who will be involved in implementing a project must be given training and resources. The certification will not be applicable to one department or section of the company, so the key process in every department should be involved so that there is no issue of coordination and cooperation from all departments. Their roles and responsibilities must be well defined.
2. *Training of employees*: The key functionaries and also others who have been identified for the project are to be provided with detailed training. The key functionaries will act as internal auditors and thus will play an important role in maintaining and improving the system. They should know about the required standards and parameters so that action can be taken to meet those standards. The rest of the employees are also to be educated about the standards and requirements.
3. *Preassessment*: This is an exercise in gap identification. The standards have been fixed and thus the desired level is known. An extension exercise must find out the present status and assess the gap between the program and the desired level of certification. Key tasks will be identified and a task list will be prepared. Timelines are to be fixed and responsibilities delegated for completing different tasks. It is better that no changes to a process be made

without involving those who are responsible for performing it. If needed the necessary changes can be made to improve the process. The gap analysis is also an opportunity to evaluate the current process for effectiveness and efficiency. There will be a need to streamline the process, record tasks, and document each step. A procedure will have to be prepared to describe the process to meet the requirements of the chosen standard.

4. *Preaudit*: This is an opportunity to have a complete drill of all departments before the real audit. It provides the organization with a mechanism to determine whether it is ready for an official audit. This must be done as it gives the chance to internally improve processes and systems. A preaudit can be conducted by a certification body or it can be done internally by designating a special team for the purpose.

Steps to Gaining Certification

Introduction

For any FSMS, organizations are certified upon completion of a satisfactory audit and a positive certification decision from a certification body (CB), also called a registrar. The registrar will have been assessed in turn and judged as competent by an accreditation body (AB). The process for certification of organizations is outlined in the flowchart (Figure 5A.1).

In order to receive a valid certificate, the organization shall select a registrar that is approved and licensed by an AB. The AB stipulates detailed requirements that a registrar shall meet in order to gain approval.

A FSMS certification can be completed within 3–6 months. The duration very much depends on the type and size of the company, the complexity of its processes, and the controls available in the company to minimize food safety hazards.

Selection of a CB

It is essential that the organization is assessed against the latest version of the chosen standard scheme and that the scheme is available throughout the certification process. The scheme should be read and understood, and a preliminary self-assessment

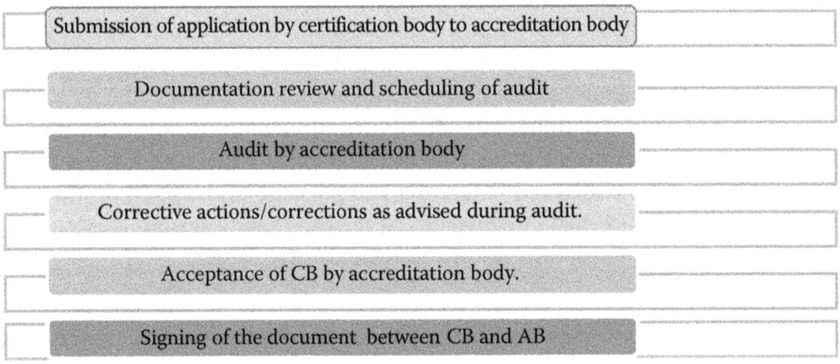
Submission of application by certification body to accreditation body
Documentation review and scheduling of audit
Audit by accreditation body
Corrective actions/corrections as advised during audit.
Acceptance of CB by accreditation body.
Signing of the document between CB and AB

FIGURE 5A.1 Flowchart of process for recognizing certification body.

will be conducted by the organization against the requirements and guidance of this scheme. Any area of non-conformity shall be addressed by the organization. Once the self-assessment has been completed and nonconformities addressed, the organization must shortlist and select a registrar.

For a list of approved registrars for certification of the FSSC 22000, the BRC Global Standard for Food Safety, and the SQF Code, refer to Appendix 1.

Certification Agreement

Quotations will be invited from the shortlisted registrars and a certification agreement finalized. The contract between the organization and the registrar will detail the agreed scope of the audit, including reference to the requirements of the food safety standard scheme. This contract shall be formulated by the registrar. It is the responsibility of the organization to ensure that adequate and accurate information is given to the registrar to enable them to select an auditor(s) with the required skills to undertake the audit. The registrar shall require completion of an official application form, signed by a duly authorized representative of the applicant.

Audit Program, Duration, and Cost

For the initial audit, the organization shall agree a mutually convenient date(s), with due consideration given to the amount of work required to meet the requirements of the scheme. The organization shall provide the registrar with appropriate information to allow them to review the application and to assess the duration and the cost of the audit.

There is a requirement for the organization to plan carefully for the audit, to have appropriate documentation for the auditor to assess, and to have appropriate staff available at all times during the on-site audit. A plan for the audit should be provided by the registrar to ensure that the food safety team and all company employees are properly prepared. The audit may be extended if staff or documentation are not available at the audit, so preparation is essential. It is important that the site is in production at the time of the audit, otherwise a further audit will be required.

The initial certification is carried out at the premises of the organization and is conducted in one or two stages, depending on the chosen standard. Note that the ISO 22000 and the SQF Code use two stages, while the BRC Global Food Standard combines the two stages. In the first stage the documentation of the food safety system is evaluated, which includes, among other things, the scope of the food safety system, the food safety hazard analyses, the PRPs, the management structure, the policy of the organization, and so on. An important objective of this audit is to assess the preparedness of the organization for the audit. Any areas of concern that could be classified as non-conformity shall be resolved before the second stage of the audit. In the second stage, the implementation and effectiveness of the food safety system is evaluated.

Corrective Actions

At the end of the audit the registrar should provide a written list of any areas that need improvement in order to gain certification; this will also be discussed at the closing meeting. Where non-conformances have been identified, these must be addressed and suitable evidence provided to the registrar for assessment within 14–28 days dependent on the level of the non-conformity (i.e., major or minor). In some circumstances it may be necessary for the auditor to return to the site to check that appropriate corrective action has been taken. The registrar will review the audit report from the auditor and corrective action documentation provided by the organization in order to make a certification decision.

Certification Granted

The registrar's audit team shall analyze and review the findings of the first- and second-stage audit and report on the assessment. Nonconformities will be pointed out as well as, where applicable, the effectiveness of the corrections and corrective action taken or planning to be taken by the organization. On the basis of this audit report and any other relevant information (e.g., the comments of the organization on the audit report) the registrar shall make a certification decision. A certificate shall only be granted if all nonconformities are resolved. In the case of minor nonconformities, the registrar may grant certification if the organization has a plan for correction and corrective action. The certificate shall be issued by the registrar typically 30–40 calendar days after the registrar has reviewed, accepted, and verified the effectiveness of the corrections and corrective actions and the plans for the corrections and corrective actions for the revealed nonconformities. The users of the certificates are advised to verify that the scope of the certificate is clearly stated and that this information is consistent with their own requirements. While the certificate is issued to the organization, it remains the property of the registrar, which controls its ownership, use, and display. The organization has the right to appeal the certification decision made by the registrar in accordance with the documented appeal-handling process of the registrar. It is a principle that the audit report is owned by the company paying for the audit and copies can be provided to other parties at the request of the owner. It is common practice to authorize the release of a copy of the report and/or certificate for customers.

Changes, Scope Extension

Once certification has been granted, any changes that may affect the fulfillment of the requirements for the certification shall immediately be communicated to the registrar. These may include changes in the products or manufacturing processes that may require extension of the scope of the certification, in the management and ownership of the organization, in the location, and so on. The registrar will then conduct a site visit to examine the consequences and determine if any audit activities are necessary. The registrar decides whether or not extension may be granted. If extension is granted, the current certificate will be superseded by a new certificate, using the same expiry dates as detailed in the original certificate.

Surveillance

The certificate expires 1–3 years after the date of issuance, depending on the standard. In the case of ISO 22000:2005, the certification audit cycle is 3 years, with a surveillance audit being conducted at least once a year. These audits shall address all scheme requirements from ISO 22000, relevant PRP documents, and the chosen food safety standard, plus use of marks and references to certification. In the case of the BRC Global Food Safety Standard and the SQF Code, the audit cycle is 6 months to 1 year, depending on the type and number of non-conformances. In case non-conformances are identified by the audit team, the registrar shall take a decision to continue, suspend, or withdraw the certificate, depending on the corrections and corrective actions of the organization.

Recertification

Before the date of expiration of the certificate, a recertification audit shall be conducted. The purpose of this audit is to confirm the continued conformity and effectiveness of the food safety system as a whole. The fulfillment of all requirements is evaluated. The audit also includes a review of the system over the whole period of certification, including previous surveillance audit reports. Identified non-conformances are dealt with as described in the surveillance audits. The registrar makes a decision about whether to renew the certification on the basis of the recertification audit, the review of the system over the whole period, and complaints received from users of the certification.

Communication with CBs

In the event that the organization becomes aware of legal proceedings with respect to product safety or legality, or in the event of a product recall, the organization shall immediately make the registrar aware of the situation. The registrar in turn shall take appropriate steps to assess the situation and any implications for the certification, and shall take any appropriate action.

CHALLENGES FACED DURING THE IMPLEMENTATION OF A FSMS

1. Management commitment is crucial because the project leader and key personnel have to be able to improve the systems to achieve certification. Proper training of personnel and budget allocation for the purpose is crucial.
2. Proper planning, realistic timelines, and fixing of roles and responsibilities are very important.
3. Ineffective gap analysis and management review, instead of helping to improve the system, should assess the actual position/situation so that improvement can be done to meet the standards. Hazard identification and analysis must be correct.
4. Incompatible cleaning and sanitation will also hinder implementation.
5. Validation and verification exercises in any process are very important.
6. The selected food supply standard should be applicable to the food supply chain, as sanction problems are inherited and cannot be solved at the last level.
7. The documentation and traceability of the supply chain is not available or it is difficult to get without taking into account the supply chain.
8. The legal and technological food supply requirements are to be integrated. PRPs, HACCP, and good practices need to be incorporated into the system.

APPENDIX 5.B

ACCREDITATION BODIES

In order to ensure that the ABs operate to a high standard of competence and probity, and that they apply the food safety standards in a consistent and equivalent manner, they shall meet the requirements of ISO/IEC 17011 and shall be a member of the International Accreditation Forum (IAF) and a signatory to the IAF ISO/IEC 17021 (QMS/EMS) Multilateral Recognition Arrangement.

The IAF is the world association of conformity assessment ABs and other bodies interested in conformity assessment in the fields of management systems, products, services, personnel, and other similar programs of conformity assessment. Its primary function is to develop a single worldwide program of conformity assessment that reduces risk for business and their customers by assuring them that accredited certificates may be relied upon. Accreditation assures users of the competence and impartiality of the accredited body. IAF members accredit certification or registration bodies that issue certificates attesting that an organization's management, products, or personnel comply with a specified standard (called a "conformity assessment").

The following ABs recognize the FSMS scheme FSSC 22000 as a certification scheme that can be certified under ISO/IEC 17021 accreditation. They will offer accreditation services to interested certification bodies that want to demonstrate that they meet the requirements of the FSSC 22000 scheme and ISO/IEC 17021:

ANSI-ASQ National Accreditation Board (ANAB) (www.anab.org)
Belgian Accreditation Organization (BELAC) (www.belaf.fgov.com)
Comite francais d'accreditation (COFRAC) (www.cofrac.fr)
Czech Accreditation Institute (CAI) (www.caLcz)
Danish Accreditation and Metrology Fund (DANAK) (www.danak.dk)
Deutsche Akkreditierungsstelle (DAkkS) (www.dakks.de)
Dutch Accreditation Council (RvA) (www.rva.nl)
Entidad Mexicana de Acreditaclon, a.c (EMA) (www.ema.org.mx)
Entidad Nacional de Acreditacion (ENAC) (www.enac.es)
Hellenic Accreditation System SA (ESYD) (www.esyd.gr)
Irish National Accreditation Board (INAB) (www.inab.ie)
Italian Accreditation Body (ACCREDIA) (www.accredia.it)
International Accreditation Forum (IAF) (www.iaf.nu)
Japan Accreditation Board (JAB) (www.jab.or.jp)
Joint Accreditation System of Australia & New Zealand (JAS-ANZ) (www. jas-anz.com.au)
National Accreditation Board for Certification Bodies India (NABCB) (www. qcin.org)
Norwegian Accreditation (NA) (www.akkretitert.no)
Organismo Argentino de Acreditacion (OAA) (www.oaa.org.ar)
Organismo de Acredltaclon Ecuatoriano (OAE) (www.oae.gob.ec)
Romanian Accreditation Association (RENAR) (www.renar.ro)

Standards Council of Canada (SCC) (www.scc.ca)
Standards Malaysia (DSM) (www.standardsmalaysia.gov.my)
Swedish Board for Accreditation and Conformity Assessment (SWEDAC) (www.swedac.se)
Swiss Accreditation Service (SAS) (www.seco.admin.ch)
Turkish Accreditation Agency (TURKAK) (www.turkak.org.tr)
United Kingdom Accreditation Service (UKAS) (www.ukas.com)

APPENDIX 5.C

CERTIFICATION AGENCIES

Certification agencies are licensed to issue accredited certificates based on global food safety standards such as the key standards FSSC 22000, BRC Global Food Standard, and SQF Code. These licenses are based on agreements between the owner/providers of the standards and the agency and require accreditation for these standards. Accreditation will be gained in conformance with the standard schemes by ABs that comply with the regulations for ABs. Licensed CBs are obliged to adhere strictly to these schemes.

Accredited CBs for the FSSC, BRC, and SQF standards are listed below (in alphabetical order):

- *FSSC*
  - 3EC International (www.3ec.cz)
  - AENOR (www.aenor.es)
  - AFNOR Certification (www.afnor.org)
  - AIB International (www.aibonline.org)
  - Applus Mexico (www.applus.com)
  - AsureQuality (www.asurequality.com)
  - Audis Corporation (www.audis.jp)
  - BMG Trada/ProSanitas Certifiering AB (www.prosanitas.se)
  - Bohemia Certification (www.bohemiacert.cz)
  - BSI Group (www.bsigroup.com)
  - Bureau de Normalisation du Quebec (www.bnq.qc.ca)
  - Bureau Veritas Certification (www.bureauveritas.com)
  - Cert ID LC (www.cert-id.com)
  - Certind S.A. (www.certind.ro)
  - Certiquality Srl. (www.certiquality.it)
  - CSQA Certificazioni Sri (www.csqa.it)
  - DNV Business Assurance (www.dnvba.nl)
  - DQS-UL Food Safety Solutions GmbH (www.dqs.de)
  - DS Certificering A/S (www.dscert.dk)
  - Eagle Food Registrations Inc. (www.eagleregistrations.com)
  - EQA Certification Mexico (www.eqamexico.com)
  - Germanischer Lloyd Certification Mexico (www.glc-mexico.com)
  - Global Standards S.c. (www.globalstd.com)

- Intertek Labtest Ltd. (www.intertek.com)
- ISACert B.V. (www.isacert.com)
- JACO (www.jaco.co.jp)
- Japanese Standards Association (www.jsa.or.jp)
- Japan Quality Assurance Organization (www.jqa.jp)
- JIA-QA Center (www.jia-page.or.jp)
- JIC Quality Assurance Ltd. (www.jicqa.co.jp)
- JMA QA Registration Center (www.jma.or.jp)
- KIWA N.V. (www.1kiwa.com)
- Lloyds Register Quality Assurance (www.lrqa.com)
- Moody International Certification LTD (www.moodyint.com)
- Nemko AS (www.nemko.com)
- NISSERT Uluslararasi Sertifikasyon (www.nissert.com)
- NKKKQA (www.nkkkqa.co.jp)
- NQF Ascertiva Group Limited (www.nqa.com)
- NSAI (www.nsai.ie)
- NSF Certification UK Ltd (www.nsf-food-europe.com)
- NSF International Strategic Registrations (www.nsf.org)
- Perry Johnson Registrars (www.pjr.com)
- ProCert SA (www.procert.ch)
- Qlip N.V. (www.qlip.nl)
- QMSCERT Ltd (www.qmscert.com)
- QSCert (www.qscert.com)
- Quality Austria (www.qualityaustria.com)
- Russian Register (www.rusregister.ru)
- SABS Commercial (Pty) Ltd (www.sabs.co.za)
- SAI Global (www.saiglobal.com)
- SGS Systems and Services Certification (www.sgs.com)
- Silliker Global Certification Services (www.silliker.com)
- SP Technical Research Institute of Sweden (www.sp.se)
- SQS Swiss Ass. for Quality & Management Systems (www.sqs.ch)
- SRAC CERT SRL (www.srac.ro)
- Test-St. Petersburg (www.test-spb.ru)
- Tuev-Thueringen (www.tuev-thueringen.de)
- Turkish Standards Institution (TSE) (www.tse.org.tr)
- TuV Austria Hellas LTD (www.tuvaustriahellas.gr)
- TuV Nord Cert (www.tuv-nord.com)
- TOV SOD America de Mexico (www.tuvmex.com.mx)
- TuV Rheinland Cert GmbH (www.tuv.com)
- TuV Sud Management Service GmbH (www.tuev-sued.de)
- Udem Ltd. (www.udemltd.com.tr)
- Union of Japanese Scientists and Engineers (www.juse-iso.jp)
- *BRC*
  - AIB International (www.aibonline.org)
  - Aus-Qual Pty Ltd (www.ausqual.com.au)

- Bureau Veritas Certification (www.bureauveritas.com)
- Cert ID LC (www.cert-id.com)
- DFA of California (www.agfoodsafety.org)
- DNV Business Assurance (www.dnvba.nl)
- DQS-UL Food Safety Solutions GmbH (www.dqs.de)
- FSNS Certification and Audit LLC (www.food-safetynet.com)
- Intertek Labtest Ltd (www.intertek.com)
- ISACert B.V. (www.isacert.com)
- Lloyds Register Quality Assurance (www.lrqa.com)
- Moody International Certification LTD (www.moodyint.com)
- NCS International Pty Ltd (www.ncsLcom.au)
- NSF Certification UK Ltd (www.nsf-food-europe.com)
- NSF International Strategic Registrations (www.nsf.org)
- Perry Johnson Registrars (www.pjr.com)
- Qlip N.V. (www.qlip.nl)
- SAI Global (www.saiglobal.com)
- SCS Global Services (www.scscertified.com)
- SGS Systems and Services Certification (www.sgs.com)
- Silliker Global Certification Services (www.silliker.com)
- TOV SOD America (www.tuvamerica.com)
- *SQF*
  - AIB International (www.aibonline.org)
  - AsureQuality Ltd (www.agriquality.co.nz)
  - Aus-Qual Pty Ltd (www.ausqual.com.au)
  - BRTUV – Avallacoes da Qualidade S.A. (www.brtuv.com.br)
  - Bureau Veritas Certification (www.bureauveritas.com)
  - Cert ID LC (www.cert-id.com)
  - Complete Integrated Certification Services (www.cicsglobal.com)
  - Det Norske Veritas- DNV (www.dnvcert.com)
  - DFA of California (www.agfoodsafety.org)
  - Eagle Food Registrations Inc. (www.eagleregistrations.com)
  - Eurofins Certification (www.eurofins.com)
  - FSNS Certification and Audit LLC (www.food-safetynet.com)
  - Global Standards S.c. (www.globalstd.com)
  - Intertek Labtest Ltd (www.intertek.com)
  - NCS International Pty Ltd (www.ncsi.com.au)
  - NSF International Strategic Registrations (www.nsf.org)
  - Perry Johnson Registrars (www.pjr.com)
  - ProCert SA (www.procert.ch)
  - SAI Global (www.saiglobal.com)
  - SCS Global Services (www.scscertified.com)
  - SGS Systems and Services Certification (www.sgs.com)
  - Silliker Global Certification Services (www.silliker.com)
  - TOV SOD America (www.tuvamerica.com)
  - Validus (www.validusservices.com)

APPENDIX 5.D

THE ISO 22000:2005 FSMS

The requirements for any organization in the food chain (ISO 22000, in short) have been specifically developed to harmonize all individual (national) standards. By integrating multiple principals, methodologies, and applications, ISO 22000 is easier to understand, apply, and recognize. It can be used by all organizations in the food supply chain, from farmers to food services to processing, transportation, storage, retail, and packaging. This makes it more efficient and effective as an entry-to-market tool than previous combinations of national standards. The ISO 22000:2005 standard is not a GFSI-benchmarked scheme. The standard, however, has become increasingly popular worldwide, especially in Asia.

The ISO 22000 intends to define the food safety management requirements for companies that need to meet and exceed food safety regulations all over the world: one standard that encompasses all the consumer and market needs. The ISO 22000 standard, officially available since September 2005, emphasizes the certification requirements for HACCP and will further contribute to the standardization and harmonization of HACCP systems worldwide. This speeds and simplifies processes without compromising other quality or safety management systems. Increasing consumer demand for healthier food for safety reasons and for ISO 22000 certification is increasing day by day. The introduction of a standard specific to food safety is extremely beneficial. It helps to improve the business, quality, and food safety system control and risk analysis. Risk analysis should be at the heart of food production companies. Food producers/processors should ask themselves the following questions:

- What can happen?
- How do I prevent it happening?
- If that goes wrong, what can I do to put it right and stop it from recurring?

The importance for the food industry is that all standards are HACCP based, otherwise they are not applicable. Companies with ISO 9001 have never had an advantage in the food industry. Now, having ISO 22000, which is a risk-assessment-based system like HACCP, means that a food company will have international recognition, and many companies will say, "I don't need to do a supplier audit because the company is achieving ISO 22000 standards." This is a turning point in standard terms. Based on the Deming plan–do–check–act cycle, ISO 22000:2005 has the same basic structure as other international management system standards such as ISO 9001 or ISO 14001. It offers a common framework for integrating different management systems. In comparison with ISO 9001, the standard is more procedure orientated than principle based. Apart from that, ISO 22000 is an industrial-specific risk management system for any type of food and marketing, which can be closely incorporated with the quality management system of ISO 9001.

The ISO 22000:2005 standard is generic (i.e., applicable to all organizations, regardless of size, which are involved in any aspect of the food supply chain and want to implement systems that consistently provide safe products), so it applies to both food manufacturing as well as food-related industries like packaging, storage, and distribution channels in public and private sectors. The standard says what should be done by an organization to manage the food safety hazards of its activities, but does not dictate how to do it. Any organization may develop its own ISO 22000 food safety system to address food safety issues arising out of its activities, product, or services. There are many requirements and subelements in ISO 22000.

The fact that food can be rendered unsafe for consumption at any level of the food supply chain, but that only the roles of food processors are normally highlighted, has piloted the need for traceability across the whole food supply chain. The need for international standards has become imperative in establishing a principle framework to monitor the path of food. ISO 22000 uses the same definition of food traceability as the CAC. At each level of the supply chain, the participant is well informed as to who their supplier is and to whom they are supplying. This methodology allows the entire path followed by the food before it reaches the shelf to be mapped. The key elements of the ISO 22000:2005 standard are

- Policy development
- Organizational development
- Planning and implementation (developing planning techniques)

ISO 22000 specifies requirements that enable a food producer/processor to

- Plan, implement, operate, maintain, and update a FSMS aimed at providing products that, according to their intended use, are safe for the consumer
- Demonstrate compliance with applicable statutory and regulatory food safety requirements
- Evaluate and assess customer requirements and demonstrate conformity with those mutually agreed upon customer requirements that relate to food safety, in order to enhance customer satisfaction
- Effectively communicate food safety issues to their suppliers, customers, and relevant interested parties in the food supply chain
- Ensure that the food producer/processor conforms to its stated food safety policy, to demonstrate such conformity to relevant interested parties
- Seek certification or registration of its FSMS by an external organization

General Requirements

- The food producer/processor is required to establish, document, implement, and maintain an effective FSMS and update it when necessary in accordance with the requirements of the ISO 22000 standard.
- The food producer/processor is required to define the scope of the FSMS, specifying the products or product categories, processes, and production sites addressed by the FSMS.
- The organization is required to
    - Ensure that food safety hazards are identified, evaluated, and controlled in such a manner that the food products do not, directly or indirectly, harm the consumer

- Communicate appropriate information throughout the food supply chain regarding safety issues related to its products
- Communicate information concerning the development, implementation, and updating of the FSMS throughout the organization, to the extent necessary to ensure the food safety required by the standard
- Evaluate the FSMS periodically, and update it when necessary, to ensure that the system reflects the organization's activities and incorporates the most recent information on the food safety hazards subject to control

DOCUMENTATION REQUIREMENTS

The FSMS documentation must include

- Documented statements of a food safety policy and related objectives
- Documented procedures and records required by this standard
- Documents needed by the organization to ensure the effective development, implementation, and updating of the FSMS

CONTROL OF DOCUMENTS

Documents and records required by the FSMS must be controlled to ensure that all proposed changes are reviewed prior to implementation to determine their effects on food safety and their impact on the FSMS. A documented procedure must be established to define the controls needed to

- Approve documents for adequacy prior to issue
- Review and update documents as necessary, and reapprove documents
- Ensure that changes and the current revision status of documents are identified
- Ensure that relevant versions of applicable documents are available at points of use
- Ensure that documents remain legible and readily identifiable
- Ensure that relevant documents of external origin are identified and their distribution controlled
- Prevent the unintended use of obsolete documents, and ensure that they are suitably identified as such if they are retained for any purpose

CONTROL OF RECORDS

Records must be established and maintained to provide evidence of conformity to requirements and evidence of the effective operation of the FSMS. Records must remain legible, readily identifiable, and retrievable. A documented procedure must be established to define the controls needed for the identification, storage, protection, retrieval, retention time, and disposition of records.

DELIVERABLES

The deliverables of ISO 22000:2005 can be summarized as follows:

- The standard can be used as the basis of any food safety management system, with or without third-party certification.
- The standard includes requirements for addressing (i.e., assessing and implementing) the food safety concerns of customers (e.g., retailers) and regulators.
- The standard includes PRPs (good hygiene practices) and HACCP.
- System management (ISO 9000 approach).
- Interactive communication along the food chain.

PREREQUISITE PROGRAMS

Any food company must have PRPs in place before setting up a HACCP system. PRPs are an essential part of the FSMS being audited, making a HACCP system effective. The company must demonstrate its commitment to food safety and meeting legal requirements. In the working document ISO 22000 these areas are part of a food company's risk assessment. It is important that food companies identify and control all hazards within the company and that the members of the management team implementing the FSMS be given the authority and responsibility to enforce these. Some food companies identify (too) many CCPs that are technically quality control points. This happens when organizations are influenced by their customers or third-party auditors.

The following PRPs *must* be considered for HACCP implementation:

- Personnel hygiene
- Hygiene of the environment and the layout of the premises
- Infrastructure, construction, and layout of the buildings including workspaces, employee facilities, the maintenance department, laboratory, and warehouse
- Supplies of utilities like water, energy, (compressed) air, steam, and others; supporting services including waste and sewage disposal
- Hygienic design and suitability of the equipment, and its accessibility
- Maintenance and preventive maintenance programs
- Management of purchased raw and packaging materials, ingredients, and chemicals such as ink and cleaning agents
- Storage of raw and packaging materials and finished goods (warehousing) and subsequent handling and transportation
- Measures to prevent cross-contamination
- Cleaning and sanitizing
- Pest control
- Traceability

According to ISO 22000:2005, the PRPs *must* be documented, and this documentation is largely dependent on the size of the company. The ISO 22000 has been designed for large companies but can also be used for small and medium-sized enterprises.

CLAUSES AND MANDATORY PROCEDURES

The main clauses of ISO 22000:2005 are

- Clause 4: Food safety management system
- Clause 5: Management responsibility
- Clause 6: Resource management
- Clause 7: Planning and realization of safe products
- Clause 8: Verification, validation, and improvement of the FSMS

ISO 22000:2005 states that the following documented procedures *must* be documented:

1. Document control (Clause 4.2.2)
2. Record control (Clause 4.2.3)
3. Handling of potentially unsafe products (Clause 7.6.5)
4. Identification and assessment of affected (end) products to determine their proper handling and a review of the corrections (Clause 7.10.1)
5. Corrective actions and the causes of detected nonconformities (Clause 7.10.2)
6. Withdrawal of products, which includes traceability (Clause 7.10.4)
7. Internal auditing (Clause 8.4.1)

PLANNING AND REALIZATION OF SAFE PRODUCTS

Clause 4 refers mainly to the HACCP system.

The ISO 22000 standard is an auditable standard and the HACCP system has to comply with the following:

- The food producer/processor must plan and develop the processes needed for the realization of safe products.
- The food producer/processor must implement, operate, and ensure the effectiveness of the planned activities and any changes to those activities. These actions include the PRPs as well as operational prerequisite programs (OPRPs).

Followings are the detailed steps that need to be taken for planning and realization of safe products:

- PRPs
- Preliminary steps to enable the hazard analysis, which include:
    - The food safety team
    - Raw materials and final product specifications and their intended use
    - Flow diagrams and descriptions of process steps and control measures

- Hazard analysis
- OPRPs
- The HACCP plan, which includes:
    - The CCPs
    - Determination of the critical limits for the CCPs
    - The monitoring system the organization shall establish and maintain for a documented procedure (nr 3) of the appropriate handling of unsafe products to ensure that the product is not further processed or released until the product has been evaluated

Note: This procedure can be combined with correction/corrective action.

The following actions will be required to be taken whenever monitoring results exceed critical limits.

- Updating preliminary information and documents specifying the PRPs and the HACCP plan
- Verification planning
- Using the traceability system

VALIDATION, VERIFICATION, AND IMPROVEMENT OF THE FSMS

The food safety team shall plan and implement the processes needed to validate control measure combinations and to verify and improve the FSMS.

The validation, verification, and improvement of the FSMS involve the following:

- Validation of control measure combinations
- Control of monitoring and measuring
- FSMS verification

INTERNAL AUDIT

The organization shall conduct internal audits at planned intervals. The responsibilities and requirements for planning and conducting audits and for reporting results and maintaining records shall be defined in a documented procedure (nr 7), including

- Improvement
- Continual improvement
- Food safety management updating

REQUIRED DOCUMENTS AND RECORDS

Next to mandatory procedures under ISO 22000:2005, the following documents and records are required:

Record Number	Clause or Subclause	Type of Record
1	5.6.1	Records of customer complaints
2	5.6.1	External communication records
3	5.8.1	Records of management reviews
4	6.2.1	Records of agreements or contracts with external experts with defined responsibilities and authorities
5	6.2.2	Records of training of competencies of personnel, and the personnel responsible for monitoring, corrections, and corrective actions of the FSMS
6	7.2.3	Records of verification and modifications
7	7.3.1	Records of relevant information to conduct the hazard analysis
8	7.3.2	Records of experience and knowledge of the food safety team
9	7.3.5.1	Records of verification results of the flow charts
10	7.4.2.1	Records of food safety hazards reasonably expected for the type of product and during processing
11	7.4.2.3	Records of the justification for and results of the determination of the acceptable level of the food safety hazard in the end product
12	7.4.3	Results of the food safety hazard assessment
13	7.4.4	Results of the assessment to categorize control measures to be managed through OPRPs or by the HACCP plan
14	7.5	Records of monitoring OPRPs
15	7.6.1/4	Records of monitoring CCPs under the HACCP plan
16	7.8	Results of verification
17	7.9	Records of traceability
18	7.10.1	Records of evaluation of nonconformities
19	7.10.1	Records of correction results (which can be combined with the evaluation record)
20	7.10.2	Records of results of corrective action
21	7.10.3.1	Records of withdrawal (recall) of the product
22	7.10.4	Records of the cause, extent, and result of a withdrawal
23	7.10.4	Records of the verification of the effectiveness of the withdrawal program (i.e., the results of a mock recall)
24	8.2	Records of validation results of OPRPs and the HACCP plan
25	8.3	Records of the results of calibration and verification
26	8.3	Records of assessment of nonconforming results of measuring equipment and resulting actions
27	8.4.1	Records of internal audits
28	8.4.3	Records of analysis of the results of verification activities
29	8.5.2	Records of system updating activities

Series of ISO 22000 Standards

ISO 22000:2005 is a series of a number of ISO documents. These standards are known as the ISO 22000 family of standards:

- ISO/TS 22002-1:2009 PRPs on food safety, Part 1: Food manufacturing:
  - ISO/TS 22002-1:2009 specifies requirements for establishing, implementing, and maintaining PRPs to assist in controlling food safety hazards.
  - ISO/TS 22002-1:2009 is applicable to all organizations, regardless of size or complexity, which are involved in the manufacturing step of the food chain and that wish to implement PRPs in such a way as to address the requirements specified in ISO 22000:2005, Clause 7.
- ISO/TS 22002-3:2011 PRPs on food safety, Part 3: Farming:
  - ISO/TS 22002-3:2011 specifies requirements and guidelines for the design, implementation, and documentation of PRPs that maintain a hygienic environment and assist in controlling food safety hazards in the food chain.
  - ISO/TS 22002-3:2011 is applicable to all organizations (including individual farms or groups of farms), regardless of size or complexity, which are involved in the farming steps of the food chain and that wish to implement PRPs in accordance with ISO 22000:2005, Clause 7.2.
- ISO/TS 22003-2007 FSMS, requirements for bodies providing audit and certification of food safety management systems: These will give harmonized guidance for the accreditation (approval) of ISO 22000 certification bodies and define the rules for auditing a FSMS as conforming to the standard.
- ISO/TS 22004: 2005 FSMS, guidance on the application of ISO 22000:2005: This was published in November 2005 and provides information to assist organizations including small and medium-sized enterprises around the world.
- ISO 22005-2007, traceability in the feed and food chain: General principles and guidance for system design.

APPENDIX 5.E

The International Accreditation Forum, Inc. (IAF) is the world association of conformity assessment ABs and other bodies interested in conformity assessment. Its primary function is to reduce risk for businesses and their customers by assuring them that accredited certificates may be relied upon. Accreditation assures users of the competence and impartiality of the accredited body. IAF members accredit certification or registration bodies that issue certificates attesting that an organization's management, products, or personnel comply with a specified standard (called a "conformity assessment").

The primary purpose of the IAF is twofold: firstly, to ensure that its AB members only accredit bodies that are competent to do the work they undertake and are not subject to conflicts of interest; secondly, to establish mutual recognition arrangements, known as multilateral recognition arrangements (MLAs), between its accreditation body members, which reduce risk to businesses and their customers by ensuring that an accredited certificate may be relied upon anywhere in the world. The MLAs contribute to the freedom of world trade by eliminating technical barriers to trade. The IAF works to find the most effective way of achieving a single system that will allow companies with an accredited conformity assessment certificate in one part of the world to have that certificate recognized elsewhere. The objective of an MLA is that it will cover all accreditation bodies in all countries in the world, thus eliminating the need for suppliers of products or services to be certified in each country where they sell their products or services. Certified once, accepted everywhere (International Accreditation Forum).

APPENDIX 5.F

INTERNATIONAL ACCREDITATION SERVICE
FOOD-SAFETY-RELATED ACCREDITATION PROGRAMS

The International Accreditation Service (IAS), an internationally recognized non-profit organization based in Southern California, provides accreditation in the food service industry. The IAS accreditation programs and services listed below encompass the diverse needs required within the food industry to enable farmers, manufacturers, distributors, and service providers to provide consumers with safe and high-quality consumable products.

1. *Inspection agencies*: Inspection agencies can be accredited to ISO/IEC Standard 17020, requirements for the operation of various types of bodies performing inspection and related standards.
2. *Product certification agencies*: Product certification agencies can be accredited to ISO/IEC Standard 17065 Conformity Assessment, requirements for bodies certifying products, processes, and services, and related standards.
3. *Management system CBs*: Management system CBs can be accredited to ISO/IEC 17021 Conformity Assessment, requirements for bodies providing audit and certification of management systems including food safety management.
4. *Personnel CBs*: Personnel CBs are accredited to ISO/IEC Standard 17024, general requirements for bodies operating certification of persons.

REFERENCES

BRC Global Standards. (2011). *BRC Global Food Safety Standard Issue 6*. BRC Bookshop, U.K. ISBN: 9780117069671. www.brcglobalstandards.com/, Accessed on March 13, 2017.
BRC Global Standards. www.brcglobalstandards.corn, Accessed on March 13, 2017.
BRC. www.brc.org.uk, Accessed on March 13, 2017.
CanadaGAP. www.canadagap.ca, Accessed on February 2, 2016.
FIA. (2012). Harmonisation of food standards in ASEAN. Food Industry Asia. https://food-industry.asia/documentdownload.axd?documentresourceid=659, Accessed on February 2, 2016.

Foundation for Food Safety Certification. http://fssc22000.com/downloads/Benefits.pdf, Accessed on February 2, 2016.

FSSC22000. www.fssc22000.com/documents/home.xml?lang=en, Accessed on March 13, 2017.

FSSC22000. (2013). Food Safety System Certification 22000. Foundation for Food Safety Certification, the Netherlands.

GAA. www.aquaculturealliance.org/, Accessed on March 13, 2017.

GFSI. (January 2011). *Guidance Document*, 6th edn. www.mygfsi.com/gfsifiles/What_is_the_GFSI_Guidance_Document_Sixth_Edition.pdf, Accessed on February 2, 2016.

GFSI. www.mygfsi.com/, Accessed on March 13, 2017.

Global Food Safety Initiative. www.mygfsi.com/, Accessed on March 13, 2017.

GLOBALG.A.P. www.globalgap.org/uk_en/, Accessed on March 13, 2017.

GRMS. www.grms.org/, Accessed on February 2, 2016.

IFS. www.ifs-nederland.nl/en/, Accessed on March 13, 2017.

International Accreditation Forum. http://www.iaf.nu/articles/About/2, Accessed on February 2, 2016.

International Accreditation Service. http://www.iasonline.org/More/2014/Food_Safety_Brochure.pdf, Accessed on February 2, 2016.

International Standard Organization. www.iso.org/home.html, Accessed on February 2, 2016.

ISO. www.iso.org, Accessed on February 2, 2016.

Prakash J. (2013). The challenges for global harmonisation of food safety norms and regulations: issues for India. Department of Food Science and Nutrition, University of Mysore, Mysore, India.

Safe Quality Food Institute. www.sqfi.com/, Accessed on February 2, 2016.

Segerson K. (1998). Mandatory vs. voluntary approaches to food safety. Food Marketing Policy Center Research Report No. 36, Department of Agricultural and Resource Economics, The University of Connecticut, Mansfield, CT.

6 Food Safety Assurance Systems

Food safety has been identified at domestic, regional, and international levels as a public health priority, as unsafe food causes illness in millions of people every year and many deaths. Key global food safety concerns include the spread of microbiological hazards; chemical food contaminants; assessments of new food technologies (such as genetically modified food); and the creation of strong food safety systems in most countries to ensure a safe global food chain. Safe food is essential for living a healthy life, but if safety is compromised it becomes the major source of ill health and diseases. According to the World Health Organization, serious outbreaks of foodborne disease have been documented on every continent in the past decade, and in many countries the rates of related illnesses are increasing significantly. It is estimated that up to one-third of the population in developed countries is affected by microbiological foodborne diseases each year, while the problem is likely to be even more widespread in developing countries. In addition, chronic dietary exposure to various chemical toxicants, which enter the food chain at different levels, can also be considered a major problem. Moreover, in developing countries there is an extra problem regarding the dilemma of food safety versus food security. Adoption of agricultural practices, such as the use of fertilizers and pesticides to increase yield, and the storage and processing of food are also increasing the chances of contamination and exposure to chemical toxicants. All these concerns have resulted in efforts being made by the food industry to create systems that can assure the consumer about the safety of food. Such efforts to create food assurance systems are a continuous process, but the last decade has seen that these efforts toward the development and implementation of different food safety management systems to assure food safety in the agriculture food chain have increased in manifold ways. As a result of this, more stringent requirements for foodstuffs have been specified as well and, as a corollary to that, so have requirement assurance systems. Usually such assurance systems cover different aspects of food safety (e.g., microbiological, chemical, and physical hazards) and the possibilities of controlling these hazards along the agri-food chain (e.g., farming, processing, and trading/distribution of food products) at different levels (e.g., company, stakeholder as government, NGO, sector association, university, or research institute levels). Food business operators (FBOs) and even other stakeholders have to familiarize themselves with Codex Alimentarius guidelines or legal requirements, knowledge of microbiological and chemical food safety, implementation of food safety management systems based on prerequisite programs, and Hazard Analysis Critical Control Points (HACCP) principles. A large number of voluntary standards are also filling the gap, as international quality assurance standards such as ISO 22000 and GLOBALG.A.P. and similar standards are gaining prominence.

Sampling and monitoring plans, inspection, food audit schedules, and storage/processing techniques have become a part of marketing strategy. The concept of risk analysis and its different parts (i.e., risk assessment, risk management, and risk communication) are used at every stage in the regulatory domain. The objective of all these activities is to enhance the confidence of FBOs and also to assure the consumer about the safety of the food (ISO/TS 22004: 2005).

The task in developing nations is more difficult and complex due to the fact that food is not available for all, and in that situation the food safety aspect is sometimes overlooked. For such nations, there is a need to examine the policies and practices designed to ensure the safety of food produced. The food safety system has many parts that are administered by different governmental organizations, with poor coordination among them. The system apparently operates as two entities: one for products destined for the export market and based largely on the requirements of importing countries, and the other with lower standards and levels of enforcement for domestic market products. The top-down approach focuses more on the end product rather than the production practices. Many events have brought the issue of overall food safety to the attention of the public, and national governments are acting positively toward addressing the deficiencies of the system. But all national governments must work in concert with provincial and local authorities to improve the infrastructure for inspecting and tracking food from farms to the end consumers to ensure a greater degree of safety in the food. Work continues around the world to establish systems to minimize health risks from farm to table, to prevent outbreaks, and to promote food safety, which also will safeguard trade and support economic development. This global issue requires a multidisciplinary and collaborative approach to developing responses effective at domestic, regional, and international levels (Knaflewska et al., 2007).

6.1   OBJECTIVES OF THE FOOD SAFETY MANAGEMENT SYSTEM

Improved food safety standards, international trade, and market integration have contributed to addressing critical issues arising from the globalization of the food supply (ISO 22000: 2005). Collaboration across the public, private, and service provider sectors is required as all stakeholders have important roles to play in improving global food safety systems and supporting better access to domestic and global food markets. As countries increase their demand for food ingredients and raw materials globally, the need to safeguard public health while promoting food security, international trade, and economic development throughout the entire value chain must be met. Good food safety management makes good business sense as well. The objectives of any food safety management system, as summarized in Figure 6.1, are as follows. A food safety management system

1. Helps in the identification of potential problems
2. Ensures that controls are in place to eliminate or limit problems
3. Helps to increase staff knowledge and can be used for staff training
4. Promotes good practices
5. Helps a business to better organize and tailor its resources and budget

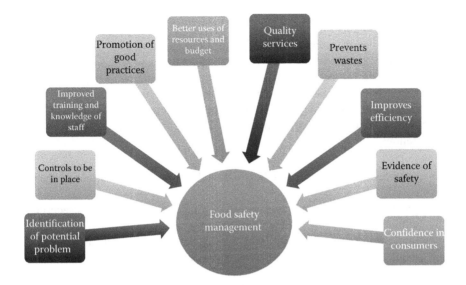

FIGURE 6.1   Objectives of Food Safety Management System.

6. Helps to provide a quality service
7. Helps to prevent wastage and spoilage
8. Improves business efficiency
9. Gives clear evidence to the enforcement authority regarding the correctness of the process
10. Creates confidence in consumers

Consumers have different and sometimes conflicting interests in relation to food; there is a need to balance those conflicting interests appropriately, though engaging regularly with the consumers. The focus should be on three main themes that need to be addressed and balanced to get the best overall outcomes for consumers:

1. The right to be protected from unacceptable levels of risk. Safety of food is the major concern.
2. The right to make choices after knowing the facts about all aspects of the food. These include safety, nutrition, and information on cost and quantity.
3. The right to the availability of the best quality supply of food in future.

Responsibilities for consumer protection are shared by all stakeholders. This includes FBOs, the food safety authority, local government, and the consumers themselves as all have important roles to play. It is the responsibility of the people producing, processing, storing, and supplying the food to ensure it is safe. The food safety authority or standards agency—whatever the case may be—has a key leadership role in making sure that concerned laws and regulations are enforced. It is a responsibility of consumers to manage the risks that they can affect relating to food. They have a right to be informed and supported in responding to those risks.

6.2   FOOD ASSURANCE STRATEGY

The food regulatory authority has to prepare a strategy to assure that food made available to consumers is safe, wholesome, and conforms to all claims and declarations made by the business. A strategy can be formulated after assessing the situation, challenges, and issues associated with providing safe food. These may vary from country to country and maybe also between regions within a country. The guidelines and points to be taken into consideration while preparing a food assurance strategy are as follows:

- Identification of areas of highest risk and targeting interventions in these areas.
- Emphasis on self-compliance and giving recognition to systems securing self-compliance.
- Increased transparency in food safety and hygiene standards, which should be made after consultations with FBOs and acceptance of practical difficulties.
- Use of wider incentives and penalties that drive compliance. This can also be done by recognizing the different drivers in different food sectors and/ or businesses.
- More emphasis on tackling persistent noncompliance with swift action on serious noncompliance. Effective and timely enforcement acts as deterrents to others. Consistent, risk-based application of controls throughout the food chain and an increased focus on their outcomes.

An important feature that distinguishes food safety and food quality is that the former is regulated by law, while the latter is demanded by those who purchase the food. Food laws and regulations are there to ensure the safety of the consumers. Quality food is demanded by manufacturers, distributors, and/or consumers, thus the safety requirements of food are minimum standards, while quality is much more than a minimum standard. Consequently, food safety assurance systems are implemented obligatorily, while quality assurance and management systems are voluntary, and thus the decision about whether or not to introduce them is to be taken by the food chain actors individually. The relationship between food legislation, safety and quality systems, official inspections, and customers is important for deciding the strategy for adopting an assurance plan. The combination of these elements forms a machinery of food safety and quality assurance and management. The implementation of a food assurance strategy has been outlined in Figure 6.2. The steps for ensuring that this food assurance strategy works well can be summarized as follows:

1. *Analysis of risk*: Risk assessment, its management, and communication are the first steps for any type of food. Thus, there will be a need to improve the overview and understanding of risk associated with food businesses. There should be clarity as to where risk lies in the food chain. Knowledge of how to handle and manage risk must be improved. A plan must be prepared so that all available resources are utilized optimally.

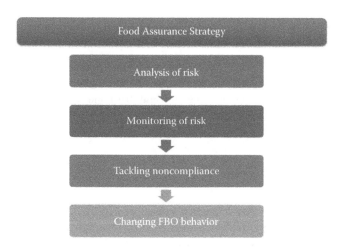

FIGURE 6.2   Implementation of Food Assurance Strategy.

2. *Monitoring risk*: Risk-based approaches must be used to monitor compliance. Efficiency must be improved and resources spared to manage key risks first. The plan should be such that they capitalize on assurance and self-compliance systems. Repeated checks should be avoided and instead these checks should be complementary to each other. Monitoring should fix priorities around removing regulatory burdens and encouraging self-regulation.

3. *Tackling noncompliance*: Government regulators have to get tougher on high-risk food products and persistently noncompliant companies. Penalties and other such provisions must be used as effectively as possible so that these can act as deterrents. Better consistency across the food chain and its regulators is a must. Regulators must be equipped with the necessary tools to take action against defaulters. Swift, effective, and reliable protection of consumers should be the objective. Optimization of the impact of enforcement will also improve the cost-effectiveness of the assurance system. Penalties should be made according to the history and extent of noncompliance and should be used effectively as deterrents.

4. *Changing FBO behavior*: Self-compliance should gain importance and messages must pass around the organization that everyone has a duty to ensure that food meets legal requirements. FBOs have to adopt an approach based on monitoring and support it through targeting and better use of compliance drivers. There should be tailored messages and guidance for targeted intervention. Regulations and laws must be modified in a way that makes it easier for FBOs to comply. Getting things right through self-compliance is more efficient than putting things right through regulation/enforcement. Systems should be such that use of financial drivers ensures that non-compliance costs more than compliance.

6.3   FOOD SAFETY MANAGEMENT SYSTEM: OVERALL MANAGEMENT OF FOOD SAFETY

There are a number of reasons for implementing Food Safety Management Systems (FSMS) in an organization. These are: certification of a product, regulatory compliance, customer protection, and gaining customer trust and satisfaction.

Any FSMS should address the following:

1. Safety
2. Risk
3. Quality

Implementing a FSMS system requires a number of steps to be considered that can be divided into the preparatory, implementation, and final phases for certification, if required. Each of the phases includes specific subjects to be considered. The first step is the selection of the food safety management standard. Sometimes, apart from other considerations, the selection of the standards is also dependent on the allocation of a budget. To select a standard as the basis for the FSMS, the following needs to be answered: will the company choose for certification of the FSMS? If so, it is recommended to select a benchmarked standard and understand the requirements of that standard. Management of food safety is basically management of the various activities taking place in an organization.

Any FSMS consists of

1. Supply chain management
2. Critical point management
3. Inspections management
4. Process control systems management
5. Preventive control management
6. Corrective actions management
7. Change management
8. Document management
9. Training management

However, no matter which FSMS is chosen, they are all similar enough that the development and the implementation will be the same.

Basically, the implementation of any FSMS will consist of the following steps:

- Introduction of the chosen standard for food safety after gap analysis (including assessment of the prerequisites)
- Senior management involvement/sensitization
- Food safety plan/HACCP implementation
- Food safety quality management system
- Training and implementation
- Internal auditing training and checklists
- Final steps to certification (if required)

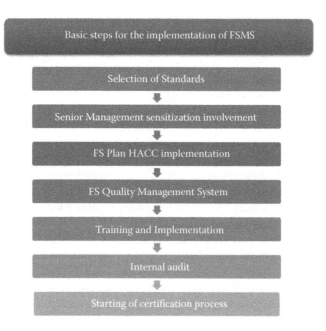

FIGURE 6.3   Basic steps for the implementation of FSMS.

Figure 6.3 gives the sequence of the implementation steps for an FSMS in any organization.

6.3.1   PLANNING FOR FSMS

Planning is the most important and crucial activity whenever any improvement or change is to be designed. In fact, half the work is done if planning has been done, and we cannot succeed without efficient planning, which takes into consideration all the aspects and the expected outcomes. As has been mentioned earlier, there are certain basic requirements, steps, and activities that are more or less the same as for overall food safety management systems, but these steps—some of which are sequential or prerequisites—have to be planned in a systematic manner. Detailed planning is a compilation of those steps in a sequence that follows a logical path.

6.3.1.1   Initiation Phase

The decision that an FSMS has to be made operational in any organization initiates discussions and deliberations about the standards and type of certifications at which they are aiming. This will set the overall objective for the FSMS. The initial discussions are very important as the aim for the FSMS needs to be defined. The FSMS may be required to meet regulatory compliances and at the same time also provide credibility to the internal food safety processes adopted by the organization. It may have to aim to keep the internal systems in place for compliances and for meeting quality parameters for internal as well as external assurances. Based on the requirements—mandatory as well as voluntary—the selection and adoption of standards is the first step.

1. *Selection of standard*: There are numerous food safety standards depending on legislative, regulatory, and customer requirements. Depending on the type of business, primary production, or processing, the appropriate category of standard is selected. This may differ from company to company, but it is always better that these standards are much higher than the mandatory requirements so that there is no worry about mandatory compliances.
2. *The type of certification of the FSMS*: The organization needs to consider whether the FSMS will need to be subject to certification. This depends on regulatory requirements and also on the decision of the organization. Apart from regulatory compliances, most companies also adopt certifications for establishing credibility among consumers.
3. *Independent gap analysis*: Once the standards have been fixed and targets identified, the organization has to conduct a gap analysis to know the areas where improvements are required. The gap analysis results should clarify whether the organization is able to implement the FSMS and can potentially achieve certification against one of the benchmarked standards within a given period of time. The organization may need to hire an experienced independent consultant with some experience in one or more of the benchmarked standards. The requirements of the standard will be followed and gaps in the system will be identified. The results can be documented in a table format. The gap analysis must include the overall conclusions about whether the company is able to implement the FSMS and achieve certification of the selected standard within a predetermined time. The conclusion of the gap analysis strongly emphasizes the quality of the infrastructure of the buildings, the processes, and the products, and to what extent levels of hygiene and prerequisite programs are in place. The gap analysis will establish the size of the project and the time frame needed to implement the FSMS. Based on the result of the gap analysis, the food safety team is composed.

6.3.1.2 Organizing for the Implementation Phase

The key to implementation is communication and training. During the implementation phase everyone operates to the procedures and collects records. There are a number of key issues that every company implementing a FSMS, which are discussed in the following sections.

6.3.1.2.1 Preparation of Organization Strategy

The implementation process starts with preparation of a comprehensive organizational strategy, with top management actively involved. The senior management has to demonstrate its commitment to the development and implementation of the FSMS. The management has to establish the foundations for the FSMS by taking the following steps:

- Formulating a checklist of customer, regulatory, statutory, and other relevant food safety requirements
- Listing the food safety requirements and developing relevant policies

- Establishing food safety objectives
- Defining the scope of the FSMS
- Preparing the plan of the FSMS using a planning matrix
- Assessing the infrastructure and work environment
- Allocating responsibility and authority
- Assessing, planning, and establishing appropriate internal and external communication channels

6.3.1.2.2   Selection and Establishment of Food Safety Team

Implementing a FSMS requires teamwork and should not depend on only one person. While building knowledge of the selected food safety standard, identify and assign a team leader. There is a wide range of quality publications and software tools designed to help the food safety team understand and implement a FSMS and have it registered. At this stage an assessment should be made by the most senior technical member of the food safety team to decide if the prerequisite standards within the facility meet the food safety requirements of the chosen standard. The nominated manager should read through the requirements of the standard and assess them for compliance, using a checklist to record their findings. The team should be multifunctional and include representatives from upper management to employees from the line. In addition to being a requirement of many of the food safety standards, multifunctional teams provide varied viewpoints. For example, a sanitation employee looks at things much differently than his production counterpart, and that difference will help create a more comprehensive system. One can also consider subteams, depending on the size of the organization. The persons involved in the purchasing of raw and packaging materials should certainly participate in the food safety team.

6.3.1.2.3   Allocation of Budget

The organization has to prepare a budget for the implementation of the FSMS. Various costs need to be considered such as training costs, possible expenses for implementing prerequisite programs (PRPs), costs of verifying and validating the HACCP system, costs for microbiological analysis, and also potential costs to verify, whether the raw material and final product comply to the standards. The budget has to include the cost of compliance with essential prerequisite programs such as infrastructure, prevention of cross-contamination, and maybe replacement of equipment. Other expenses like consultant fees, training, and software are also to be included.

6.3.1.2.4   Preparation of a Detailed Plan

Planning is vital for establishing a food safety management system. Planning has to be specific, measurable, achievable, and realistic and also should have a time frame. The planning includes a detailed plan of all the activities to be carried out. The planning should include various stages of training preceding its implementation, such as training on basic hygienic measures and PRPs followed by implementation, training on HACCP followed by implementation, and training on internal auditing followed by test audits. In the planning, considerable time has to be allocated to the implementation of the prerequisite programs. Key procedures such as documentation control and nonconformance are recommended to be implemented from a very early stage.

Depending on the size of the company, the infrastructure, the initial level of PRPs in place, and the number of people, one has to consider a minimum of 1 year for implementing a FSMS. Validation and verification are essential elements and are incorporated into the planning. Various types of software can be used for the planning, from basic Excel sheets to more advanced programs.

6.3.1.2.5   Training and Involvement of Staff

Training of staff is vital. The implementation of a food safety program will change the food safety culture in the organization. Food safety is a people-oriented system and should not be implemented in a top-down approach but instead largely with the involvement of the staff. Input of staff is therefore of vital importance. Staff not only need to be trained on basic hygienic measures and PRPs, but also on the potential hazards and risks associated with the type of product manufactured. Input of ideas and suggestions from staff is of importance, and staff need to be valued and recognized for their input.

Basic training should be given to all staff and includes

1. Job/task performance
2. Company safety and quality policies and procedures
3. Good manufacturing practices (GMPs)
4. Cleaning procedures
5. HACCP
6. Biosecurity and food defense
7. Product quality
8. Chemical control
9. Hazard communication
10. Foodborne pathogens
11. Emergency preparedness
12. Employee safety
13. Safety regulatory requirements/quality regulatory requirements

The food safety team should receive extra training on

1. Internal audit training
2. HACCP training

The management should introduce production meetings and have clear agendas and action points. Ideas and suggestions related to food safety initiated by staff should not be neglected by management. Failing or ignoring suggestions related to improving food safety in the company is deemed to be a failure to implement the FSMS.

As mentioned earlier, implementation of the FSMS is a people-oriented process that requires training of staff, and this often requires changing the mind-set of the staff.

Training may not achieve the required results or outputs as initially anticipated. It is therefore important to monitor the effectiveness of the training. The effectiveness of training can be monitored by having those persons who have been subject to training complete basic exams or questionnaires. Effectiveness can also be monitored by conducting interviews. It may be required to conduct refresher courses, depending on the subject.

6.3.1.3   Implementation and Execution Phase

The execution/implementation phase ensures that the project management plan's deliverables are executed accordingly. This phase involves proper allocation, coordination, and management of human resources and any other resources such as material and budgets. The output of this phase is the project deliverables.

6.3.1.3.1   Development of the Documentation

Depending on the type of standard, and in accordance with the size of the company, the appropriate documentation is developed.

Documentation is created to the extent that it outlines the company's intention to operate in a hygienic and safe manner. It outlines why one is in business, how one is applying the management system, and how the business is operated.

It is highly recommended to follow the ISO methodology for developing procedures and instructions, records or checklists, flow charts, and other documents such as policies according to ISO requirements. Develop standard templates for the documents; a consistent format will also improve the presentation of the documents and the effectiveness of their usage.

Provide the documents with a title or name, date of issue, revision date, and a numbering or reference system that is easy to understand, implement, and maintain. Avoid "floating" documents in the organization: these are documents without a title, date of issue, and reference number. Therefore, one of the first procedures to be developed is the control of documents and records. The procedure of document control and control of records describes how documents are developed or initiated, approved, changed, maintained, and used. In conclusion, develop and adopt a numbering system in order to trace documents from an early stage. This will reduce extra and unnecessary work. It is recommended to apply the FSMS documentation system for at least 3 months prior to conducting a possible certification. If a number of these previously mentioned and discussed topics are not or are insufficiently considered, they may become pitfalls in the implementation of a FSMS. This may be the case if no budget has been allocated, if there is no management commitment, or if there is overly optimistic planning.

6.3.1.3.2   Key Performance Indicators Related to Food Safety

A comprehensive food safety system consists of GMPs, sanitation practices, regulatory compliances, quality control, and HACCP. All standards require management review meetings. In such review meetings the emphasis is on continual improvement.

It is important that key performance indicators (KPIs) related to food safety are identified, implemented, and maintained. Table 6.1 gives a few examples of KPIs and details areas of concern for the safety of food.

TABLE 6.1

Key Performance Indicators in Food Safety—Few Examples

S. No	KPI	Issue/Area of Concern
1	Consumer complaints	Contamination, adulteration safety and quality issues
2	Glass breakage	Process facilities
3	Waste material	Stopping of production line
4	High microbial count	Raw material quality, cross-contamination, delays, processing, and environment
5	Training requirements	Technical competence and training in handling of equipment process

Examples of KPIs are as follows:

- Consumer complaints related to both food safety issues and quality issues.
- Waste materials (e.g., raw material, packaging in process, final product), which may indicate (excessive) stoppages in the production line. High frequency of stoppages may introduce contamination, so corrections may be required.
- The weight of broken glass at the filling line or at the (de)palletizers.
- The levels of the microbiological count and/or contamination of raw materials and semimanufactured products in the processing line and final product.
- The levels of verification results of the microbiological count of cleaning and sanitation. These verification results may be based on quick tests or classical methods.
- Training hours of staff related to food safety. Subjects are identified and each requires 1 hour or 2 hours of training for a staff member depending on its type. These planned training man-hours can be planned over a certain period of time. Actual total training hours are followed and expressed in terms of a percentage of completed training hours.

KPIs are set on a yearly basis and revised annually. It is the task of the food safety team leader to present the results in a trend analysis and clearly indicate continual improvement.

Management has committed itself to implementing the food safety management system and would also like to see a return of investment (ROI). Therefore, if the trend analyses can be expressed in monetary terms then they should be, so that ROI can be calculated. Results of the KPIs, presented in conclusive trend analyses, are input to the management review meetings and ROI is the major decision-making criterion.

6.3.1.4   Monitoring, Controlling, and Evaluation Phase

Monitoring and controlling consists of those processes performed to observe project execution so that potential problems can be identified in a timely manner and corrective action can be taken, when necessary, to control the execution of the project. The key benefit is that performance is observed and measured regularly to identify variances from the project management plan. This is essential for knowing and evaluating if we are working as per the plan.

Monitoring and controlling includes

- Measuring the ongoing food safety activities (i.e., "where we are")
- Monitoring the variables (e.g., cost, effort, scope, results, etc.) against the project management plan and the project performance baseline (i.e., "where we should be")
- Identifying corrective actions to address issues and risks properly (i.e., "how we can get on track again")
- Influencing the factors that could circumvent integrated change control so only approved changes are implemented

Fulfillment and implementation of these tasks can be achieved by applying specific methods and instruments of control. The following methods of control can be applied:

- Investment analysis
- Cost–benefit analysis
- Value benefit analysis
- Expert surveys
- Risk-profile analysis
- Milestone trend analysis
- Target/actual comparison

Control is that element of a food safety process that keeps it on track, on time, and within budget. Control begins early with planning and ends with a postimplementation review, having a thorough involvement with each step in the process. The FSMS may be audited or reviewed while the implementation is in progress. Formal audits are generally risk- or compliance-based and management will set the objectives of the audit. An examination may include a comparison of approved food safety management processes with how it is actually being managed in the factory/organization.

6.3.2   Auditing of FSMS

An audit is essential to identify the deficiencies of a system so that corrective actions can be taken. Norms and standards that have been fixed and adopted at the preliminary stage by management are the main criteria when doing an audit. During the audit it is assessed whether there is compliance with those norms or standards.

There are three types of audit:

1. First-party audit: A first-party audit is carried out by the organization itself (e.g., one department audits another department, or one company audits another company in the same organization). This is probably the most important audit because it gives the company an opportunity to look into their own systems, procedures, and activities to ascertain whether they are adequate and whether or not they are being complied with.
2. Second-party audit: A second-party audit can be external and carried out by a supplier, or extrinsic and carried out by a customer on a company.
3. Third-party audit: A third-party audit is an audit carried out on a company by an independent organization (e.g., at the request of a group of customers, but also for the purpose of certification).

Audits can be conducted in different ways and are thus named differently as per the procedure followed. There can be adequacy audits, system audits, desktop audits, or management audits, which are usually office audits. During these audits, the auditor verifies whether a documented quality management system exists and meets the requirements of the applicable frame of reference. A compliance audit can also be carried out at the auditor's office or at the company. This is done to verify whether the documented quality system or norm/standard is being fully implemented in the company. Normally the work of the audit is divided up; the lead auditor is responsible for the preparation part and then divides the work between the auditors. Studying the applicable standards and other relevant documents is the responsibility of each auditor (ISO 9001: 2008).

There are various strategies for executing an audit:

- Trace forwards, from raw material to end product
- Trace backwards, from end product to raw material
- Trace backwards, from critical control point (CCP) to raw material

6.3.2.1 Steps in Conducting an Audit

An audit is a systematic process that needs to be planned and executed in a systematic fashion. The following steps can ensure an effective audit:

1. Planning and preparation
2. Conducting the audit
3. Analyzing the results
4. Taking corrective actions
5. Verifying the audit objectives

6.4 FOOD SAFETY AUDIT

A food safety audit focuses on gathering information about a food business to identify any areas of potential improvement in the business's food safety processes and systems. It also identifies areas of the business that are deficient and the appropriate action needed to correct these deficiencies.

The objectives of this type of safety audit are to inform the company

- How well it is performing in food safety
- Whether managers and others are meeting the standards that the company has set for itself
- Whether the company is complying with the food safety laws that affect its business, with the view that the company will make any improvements identified as necessary from this information

An ideal food audit system must have three components:

1. Assessment and validation process with approval criteria
2. Audit methodology/process
3. Management system to respond to audit findings and monitor the efficacy of audits and approved food safety auditors

Self-compliance is being promoted everywhere and in all types of activities, and the same is true of food safety. Thus most of the laws provide that the responsibility lies with food businesses to implement preventive food safety programs. However, even governments cannot abdicate their responsibility, even when the emphasis is on self-compliance. Thus there is a shared responsibility for food safety between food businesses and government. The auditor's role is to carry out audits of the food safety programs and to assess the compliance of businesses with food safety program requirements and the requirements of the food safety standards, and then to report the outcomes of the audits and assessments to the enforcement agency. It is then the responsibility of the enforcement agency to implement appropriate enforcement measures when a food business's food safety program is not effective at producing safe food. An audit is a systematic and, wherever possible, independent examination to determine whether activities and related results conform to planned arrangements and whether these arrangements are implemented effectively and are suitable for achieving the organization's policy and objectives. The health and safety management audit should be a structured process of collecting independent information on the efficiency, effectiveness, and reliability of the system and drawing up plans for corrective action. Auditing examines each stage in the food system by measuring compliance with the controls the organization has developed, with the ultimate aim of assessing their effectiveness and their validity for the future (ANZFA).

6.5 TYPES OF FOOD SAFETY AUDITS

Audits can be classified based on their activity, frequency, purpose, and objectives. As mentioned earlier, an audit can be first-party, second-party, or third-party based on the connection of the person to the organization upon which the audit is being done. Similarly, based on purpose, an audit can be called a raw material audit, financial audit, outcome audit, and so on. The frequency of an audit can be monthly, quarterly, or annually, as per requirements. There can be formal, informal, or ad hoc audits. Normally, food safety audits can be of the following types:

1. *Audit of a food safety plan*: An audit of a food safety plan is the review of such a plan at the end of the year. Sometimes the preparation of a food safety plan and its annual review is mandated in the regulations as well. The audit consists of two parts. The first is intended to provide a simple overview of progress in terms of time. The second is intended to expand on the information provided in the first part by giving reasons as to why any missed deadlines were not met, detailing any benefits gained by the activities undertaken in the time period covered by the plan, and including any other relevant information that will assist in drawing up the plan for the next 12 months. The audit looks into the progress of the existing plan, as well as the contents and format of the plan.
2. *"Walk around" audit*: A walk around audit determines whether the food safety policies of the company are being properly implemented and identifies areas in which policy effectiveness needs to be improved.
3. *Food safety management audit*: One of the main problems with food safety audits of the walk around audits is that they tend to examine food safety problems based on their symptoms rather than their causes. They rarely focus entirely on the management of food safety system.
   Food safety management audits look into the following questions:
   a. Does the company have adequate procedures for identifying specific food safety requirements that apply to its undertakings?
   b. Are the procedures followed and are responsibilities set out clearly and understood?
   c. Does the company's food safety policy documentation include adequate procedures for identifying hazards that exist, and for assessing regularly the risks in order to identify the measures needed to avoid harm?
   d. Are adequate risk assessment procedures also set out for the hazards of the products and/or services supplied by the company, in order to identify the measures needed to avoid risk of harm to customers, end users, and members of the public?
   e. Are the procedures in c. and d. followed and are responsibilities set out clearly and understood?
   f. Does the company have adequate procedures for setting, reviewing, and revising as necessary its food safety standards for meeting specific food safety requirements?
   g. Do the procedures for setting company standards include the identification of measurable targets that can be audited to monitor the level of compliance with company standards?
   h. Are the procedures in f. and g. followed and are responsibilities set out clearly and understood?

    i. Does the company have adequate procedures for planning, implementing, controlling, monitoring, and reviewing the measures identified in c. and d.?

    j. Does the company have adequate procedures for carrying out food safety audits to check that the procedures in i. are followed and that the measures in c. and d. are effective?

4. *Process safety audits*: A process safety audit is a type of self-evaluation audit that aims at

    a. Gathering all relevant documentation covering process safety management requirements at a specific facility

    b. Determining the program's implementation and effectiveness by following up on their application to one or more selected processes

5. *Product safety audits*: Product safety audits are important in the product design and development stages. They are designed to ensure that the company has adequately protected the user of a product from hazards that did not exist. This type of audit aims to

    a. Identify and classify hazards associated with the product (i.e., catastrophic, critical, occasional, remote, or improbable)

    b. Develop a hazard risk index and priority setting

6.6 INTERNAL AUDITING

Internal audits are a valuable tool for assessing the effectiveness of GMPs, HACCP, or any other food safety system. They help to identify the strengths and weaknesses of an existing food safety system and discover areas for improvement before an external audit occurs. Thus, one can implement corrective actions and prevent reoccurrences. Internal audits need to be more strict and rigorous than external audits, and management commitment is essential to ensure that the audits are conducted regularly and remain effective. The frequency of internal audits depends on each plant and process. One can conduct an initial internal audit of an entire system a few months after GMPs/a HACCP system has been implemented. This will help in assessing the progress of the implementation of the food safety system. The gap should be such that it allows sufficient time to generate records, identify deficiencies, take corrective actions, and address problems and nonconformances. Recurrent deviations and changes in the food safety system in the organization may require scheduling an internal audit. It is always better to have an internal audit before a certification/recognition audit so that such an audit goes smoothly and unnecessary repetitions and cost escalations can be avoided.

The success of any internal audit is dependent on the selection of a competent and experienced auditor. This is crucial because the selection of the right person will help to achieve the desired results, will be cost-effective, and more importantly will identify areas for business opportunities. The work of the auditor may be confined to food safety issues and mandatory compliances, but if handled properly and intelligently it can throw up opportunities and ideas for business improvement.

To ensure internal audits are as unbiased as possible, the person who audits a process should not be involved in the related areas, departments, or activities. Auditors must be impartial and objective and should have no interests in the process being audited. Internal audits have two main components:

1. *The systems audit*: This consists of assessing documents for completeness and checking that all GMPs/HACCP requirements are addressed. For GMPs, evaluate all components, including the sourcing and delivery of ingredients, packaging materials, shipping and receiving, and so on.
2. *The verification audit*: This basically involves on-site observations and interviews. After reviewing the documentation, check that programs, policies, and procedures are implemented as written. Ensure they are current and reflect the operation. Review records for completeness and accuracy. Visually assess elements such as employee activities; personnel practices; design, condition, and maintenance of the internal structures (e.g., floors, walls, ceilings, and light fixtures), food contact surfaces, and equipment; and so on. During the on-site audit, the auditor needs to interview employees to evaluate their knowledge of food safety and the GMPs/HACCP systems. Ask them about the activities they perform and the monitoring procedures they follow.

APPENDIX 6.A

PREPARATION OF A FOOD SAFETY PLAN

Every organization needs to prepare a food safety plan so as to ensure that it is taking sufficient measures to address the issue of food safety, and also that it is complying with all the relevant regulations. This starts with the adaptation of any food safety management system. Development of a food safety management system like HACCP requires going through a series of steps, including competing a gap assessment and writing prerequisite programs (ISO/TS 22003: 2007).

Before any plan can be prepared, gap assessment studies are conducted. A gap assessment is an on-site evaluation of the current food safety system in any facility. The gap assessment will indicate the strengths and weaknesses in the plant and provide the basis for a work plan to improve the food safety system. Based on gap analysis, prerequisite programs are to be decided. Prior to developing a HACCP plan, a food processor must develop written prerequisite programs that meet all the regulatory and program requirements. GMPs are decided based on the gaps in food safety.

A HACCP plan is a food safety system based on systematic and preventive methods for ensuring food safety. A food safety plan, which is basically a HACCP plan, is a set of written procedures that will help eliminate, prevent, or reduce food safety hazards. Food safety plans include the procedures to be followed right from the receiving/storage stage when the food enters the premises until the point where it is served to or purchased by customers. Under food safety regulations, every operator of a food service establishment and food premises where carcasses are handled or where food is processed or prepared must develop, maintain, and follow a food safety plan to ensure that a health hazard does not occur in the facility. A food safety plan must be completed and approved before a license/permit/approval is granted.

This basically addresses the issues that can lead to foodborne illnesses, including improper cooling and cold storage, inadequate processing advanced preparation, cross-contamination, and inadequate cooking. Food safety plans focus on the critical steps within the preparation of food to prevent these practices from occurring.

STEPS FOR THE PREPARATION OF A FOOD SAFETY PLAN

1. *Identification of potentially hazardous foods*: Potentially hazardous foods are those that are capable of supporting the growth of disease-causing microorganisms or the production of toxins. These are usually foods that are considered perishable. Mainly, these are
   a. All foods of animal origin (e.g., meat, fish, dairy, eggs, etc.)
   b. Foods of plant origin (e.g., vegetables, fruits, etc., that have been cut or cooked)
   c. All cooked foods
2. *Identification of CCPs in the facility*: A CCP is a step in the process where a food safety hazard can be controlled. These control points normally occur in the following steps of any processing facility.

Receiving ⟹ Preparation ⟹ Cooling ⟹ Storage ⟹

Cooking ⟹ Reheating ⟹ Hot holding

As all the sequences may not be critical in all processes, it is not important how a process chart follows these sequences.

3. *Fixing the critical limits as per the food safety standard*: A critical limit is a measurable standard or limit that must be met to control the food safety hazard at a CCP. Mainly, these are
   a. Cold storage temperature of 4°C or less
   b. Final cooking temperature of 74°C
   c. Hot holding temperature of 60°C or more
   d. Cooling food from 60°C to 20°C within 2 hours and 20°C to 4°C within 4 hours
4. *Mentoring actions/steps*: Mentoring is essential to check if the critical limits or standards fixed by facilities have been addressed. Monitoring can include measuring an internal temperature, visually assessing food, or observing practices. All monitoring results are to be recorded.
5. *Corrective measures*: If any deviations are observed during monitoring, corrective measures need to be taken. The food may require reprocessing or have to be discarded if it has spoiled.

Sample food safety plan: Below is an example of a general food safety plan that outlines typical steps in the food preparation process. All processes may not follow this template exactly, as all plans are specific to the plant or facility.

Preparation Step	Potential Hazards	Food Safety Standards	Monitoring Actions	Corrective Measures
Receiving	Substandard raw material	Food is obtained from approved sources.	Verify with supplier.	Return substandard and specified food to the supplier.
	Contamination of food	Refrigerated food temperature is 4°C or less upon receipt and subsequently.	Check temperature of food and record.	
	Pest infestation	Food is free of pests.	Visual inspection of food.	
	Damaged packaging	Packaging is undamaged.	Inspection of packaging.	
Storage	Cross-contamination	Perishable food is stored at 4°C or less.	Check temperature and record.	Adjust temperature setting or service the unit.
	Microbial growth	Store frozen food at 18°C or less.	Check temperature and record.	Move food to alternate storage unit; discard spoiled food.
Preparation of food	Contamination of food	Sanitize food contact surfaces and equipment prior to use.	Verify proper sanitizer concentration with test strips; observe practices.	Modify practices; discard contaminated food.
Processing	Survival of pathogens	Cook food to an internal temperature as desired/prescribed; hold food at a high temperature.	Check internal temperature using a probe thermometer in the thickest part of the food.	Adjust temperature settings; continue cooking until the required internal temperature is reached; discard spoiled food.
Cooling	Microbial growth	Cool foods from 60°C to 20°C within 2 hours, then from 20°C to 4°C within 4 hours; total cooling time should be 6 hours or less.	Check internal temperature of the food using a probe thermometer at various times during cooling; use a timer to ensure that food is cooled within the appropriate timeframe.	Discard food if cooling times and temperatures are not met.
Reheating	Survival of pathogens	Reheat foods to 74°C within 2 hours.	Check internal temperature using a probe thermometer in the thickest part of the food.	Continue cooking until the required internal food temperature is reached; discard spoiled food.

APPENDIX 6.B

INTERNAL FOOD SAFETY AUDIT

Internal food safety audits are important as they give opportunities to make improvements without any legal implications. It is appropriate that such audits be done by someone not directly related to the work, so that an outsider's perspective can be incorporated and the audit can be impartial and not suffer from prejudices. For a thorough checkup, especially to identify areas for improvement, there is a need to have separate audits for separate activities (e.g., for the GMPs, Good Agricultural Practices (GAP), and Standard Operating Procedures (SOPs) of sanitation, hygiene, HACCP, transportation, storage, human resource development, etc.). The following is a sample of the GMPs that can be developed for other activities as per the scale and objectives of the organization.

GMPs: Personnel

	Compliant	
Critical Items	**Yes**	**No**
1. Employees are well trained in what they do.		
2. While handling food products, employees wear the proper hair coverings, beard coverings, disposable gloves, and clean uniforms.		
3. Employees do not have any illnesses, infections, or injuries (e.g., boils, cuts) that can contaminate foods in the production area.		
4. Employees wash and sanitize their hands after each visit to the toilet.		
5. Employees maintain clean personal habits.		
6. Traffic within the plant is controlled to prevent contamination of the production area.		

Total Checked:

GMPs: Building and Facilities: Plant and Grounds

	Compliant	
Critical Items	**Yes**	**No**
1. Area around plant is clear of litter, weeds, grass, and brush.		
2. There is no standing water on the grounds (which also attracts pests).		
3. Floors, walls, ceilings, windows, and screens are properly maintained and cleaned.		
4. Production area doors and windows to the outside have fine mesh screens to keep out insects.		
5. The doors in the plant can be properly sealed.		
6. All holes and cracks have been filled in so as not to provide hiding places or entry points for pests.		
7. There is no evidence of the presence of domestic animals such as cats and dogs.		
8. Restrooms are cleaned regularly.		
9. Hand-washing facilities are furnished with paper or air hand dryers and soap.		
10. There are no leaks in the roof, skylights, windows, screens, or overhead piping.		
11. Overhead lights are covered with shields to prevent contamination of products by broken glass in case the lamps burst.		

Total Checked:

GMPs: Building and Facilities: Sanitation Operators: Pest Control

	Compliant	
Critical Items	**Yes**	**No**

1. Professional pest control services are used.
2. The pest control operator is checked regularly.
3. There is documentation of the chemicals being used.
4. There is no evidence of mites, weevils, or roaches apparent in the plant.
5. Fumigation is used safely.
6. Pest control logs and documentation are readily available.
7. Pesticides and application equipment is stored safely.
8. Products are sorted on pallets and kept at least 18 inches away from the walls.
9. The facility is well maintained.

Total Checked:

GMPs: Building and Facilities: Sanitary Facilities and Controls

	Compliant	
Critical Items	**Yes**	**No**

1. Trash, debris, and clutter are picked up both inside and outside the plant regularly.
2. All sanitation chemicals used in the plant are approved.
3. Employees eat, drink, and use tobacco products only in designated areas, and not in the production area or warehouse.
4. Old rodent excreta has been cleaned up.
5. Garbage is quickly removed and dumped in appropriate bins.
6. Garbage is kept covered.
7. The water used in the firm comes from an approved source (either the municipal supply or a tested private source).
8. Facilities have backflow and vacuum breaker valves to prevent contamination of the water supply.
9. There is no standing water around the firm (particularly in the production area, warehouse, and pack-off area).

Total Checked:

GMPs: Equipment

	Compliant	
Critical Items	**Yes**	**No**

1. All equipment that comes into contact with food is cleaned and sanitized as often as necessary to prevent contamination of the product.
2. The equipment is designed or otherwise suitable for use in a food plant.
3. There is no buildup of food or other material on the equipment.
4. There is no buildup or seepage of cleaning solvents or lubricants onto the equipment, which can contaminate foods.
5. No equipment is hard to disassemble for cleanup and inspection.
6. There is no "dead space" in or around the machinery where food and other debris can collect as a nest for insects and bacteria.
7. The surfaces of the equipment can be sanitized easily.

Total Checked:

GMPs: Production and Process Control

	Compliant	
Critical Items	**Yes**	**No**

1. Products are stored on a first-in, first-out basis to reduce the possibility of contamination through spoilage.
2. All incoming products are dated to ensure a proper rotation of stock and for internal tracking purposes.
3. Items are not overstocked.
4. Incoming vehicles are inspected.
5. Dusty, faded, or discolored containers are checked regularly.
6. All products spoiled by damage, insects, rodents, or other causes are stored in a designated "quarantine area" to prevent contact with safe products.
7. Quarantined items are disposed of quickly to prevent the development of pest breeding places.
8. All incoming materials are inspected for damage or contamination so that they can be rejected if necessary.
9. All unused materials are properly resealed to prevent contamination.
10. All food materials are stored in a safe manner (food-related items should not be stored with non-food-related items; materials should be stacked so that vents and blowers are not blocked; stacks of materials should be kept orderly for safety purposes).
11. There is an effective recall procedure set up.

Total Checked:

APPENDIX 6.C

Food Safety Audit Checklist (Sample)

The purpose of this audit is to conduct an assessment of the hygiene standards in all aspects of the food handling procedures carried out in the facility and to ensure that all relevant corrective actions are carried out and documented.

Date of Audit: _____

Name of Auditor: _____

Instructions
1. An internal food safety audit must be done at least twice per year.
2. If necessary, an audit for each food service area within the organization must be done.
3. The audit consists of two types of review:
   a. **A desktop audit**: A review of the documentation and records used as part of the organization's food safety program.
   b. **An on-site audit**: An audit of the practices and procedures being carried out during the production and service of food. During the audit, speak with staff/volunteers to gauge what is actually happening, day to day.
4. Record "NR" for questions that are not relevant.
5. On completion of the audit, develop an action plan to ensure any noncon-formances are dealt with immediately and appropriately. Ensure a comple-tion date is entered into the audit to document that corrective action has been carried out.
6. Ration and file all audits.

INTERNAL FOOD SAFETY AUDIT CHECKLIST

DATE: [   ]   **CHECKED BY:** [   ]   **AREA:** [   ]

(1) Does not address the issue   (2) Needs improvement   (3) Good   (4) Not applicable

Item	1, 2, 3, 4	Corrective Action
Supplier Program		
1. The approved suppliers list is up to date.		
2. All commercial suppliers have provided the organization with an up-to-date Food Authority License, food safety and/or HACCP certification information.		
3. All new commercial suppliers of potentially hazardous foods have trained persons for handling them.		
Receipt		
4. Staff are aware of the food safety issues in accepting receipt of incoming products.		
5. Staff are aware of the main food safety factors when inspecting a food delivery vehicle.		

Item	1, 2, 3, 4	Corrective Action

6. All products are listed on the approved suppliers list.

7. Specifications are available for all products.

8. Refrigerated and frozen products' temperatures are monitored upon receipt.

9. Upon receipt the products are stored quickly in their appropriate storage areas.

10. Food delivery vehicles are inspected (monitored regularly) before receiving goods.

Labeling and Traceability

11. All perishable items in storage are clearly labeled with a name, date of purchase, and use-by date.

12. All pre-prepared foods and works in progress in storage are clearly labeled.

Storage

13. All storage areas are neat and tidy, with food products stored off the ground and not in contact with wall surfaces.

14. All foods in storage containers are covered and labeled with the name of the product and date of receipt.

15. All packaging is in good condition.

16. Food is stored on a rotation use-by-date basis.

17. There is sufficient storage space.

18. There is a dedicated holding area for foods on hold or involved in recall.

19. The temperature of the storage areas is operating in the correct range.

20. Foods are stored to prevent cross-contamination from raw to cooked products in storage areas.

21. Equipment and door seals are in good order.

22. Chemicals and cleaning products are stored away from food storage areas.

23. Storage areas are free of evidence of pests.

24. Refrigeration appliances are calibrated on a regular basis (at least once every six months).

Cleaning

25. The cleaning schedule is displayed and is being followed.

26. There are adequate equipment and facilities to undertake cleaning effectively.

27. Sanitizers for work surfaces are readily available for use during food preparation.

28. Verification of cleaning effectiveness is regularly conducted.

29. Cleaning chemicals are stored separately from food areas.

Maintenance of Premises and Equipment

30. Equipment is in good repair and facilities are clean.

31. There is sufficient and well-maintained lighting, ventilation, and drainage.

Item	1, 2, 3, 4	Corrective Action
32. All fittings are free from cracks and crevices and in good condition.		
33. Food services equipment is free from cracks and chips.		
34. All fixed temperature monitoring gauges have been calibrated in at least the past six months.		
35. All probe thermometers have been calibrated monthly.		
36. The premises are in good repair, with clean drains, no peeling paint, no holes or gaps where pests might enter, and so on.		
37. There are building and equipment maintenance programs and they are being followed.		

Pest Control

38. There is no evidence of pest or rodent activity.		
39. Records are kept of pest control visits and the treatments administered.		
40. External openings are adequately sealed to prevent entry of pests.		

Waste

41. Waste is removed regularly.		
42. Waste disposal bins are distinguishable from food storage bins.		
43. Waste containers are covered, kept clean, and emptied after each work period.		
44. The refuse storage area is separated from the food preparation areas.		

Personal Hygiene

45. Daily hygiene practices are monitored.		
46. There are sufficient hand-washing facilities installed in all food handling areas.		
47. Food handlers wash their hands as often as necessary.		
48. Food handlers use gloves appropriately and correctly.		
49. All jewelry including watches is removed prior to commencing direct food handling.		
50. There is no evidence of eating or smoking in food preparation areas.		
51. Staff are aware of food safety practices and their responsibilities.		
52. Staff are trained in food hygiene.		

Training and Induction

53. Staff training is up to date and recorded on the food safety training register.		
54. All staff have appropriate skills and knowledge in food hygiene.		
55. Staff training records are up-to-date.		

Product Recall

56. Recall procedures have been implemented as per the procedures.		

Item	1, 2, 3, 4	Corrective Action

Food Preparation

57. Where a chemical sanitizer is used there are records to show levels are maintained.
58. Correct use of equipment/utensils prevents cross-contamination.
59. Work surfaces, utensils, and equipment are clean.
60. Chemicals are stored in a manner to prevent contamination.
61. The risk of foreign objects (e.g., physical items) is controlled to prevent contamination.
62. Staff are wearing appropriate protective clothing.
63. Staff are following good hygiene practices.
64. There are adequate hand-washing and drying facilities for staff.

Thawing

65. Raw products are thawed separately from cooked products to prevent cross-contamination.
66. Products being thawed are covered and/or wrapped and labeled.

Cooking

67. Cooking times and temperatures are satisfactory and monitored by staff.
68. All necessary steps are taken to prevent the likelihood of food being contaminated with microorganisms or allergens during the cooking process.
69. The flow of food is such that there is no likelihood of cross-contamination from raw, unprocessed food to ready-to-eat food.
70. The risk of postcooking cross-contamination is controlled.

Cooling

71. Records of temperature monitoring for all refrigerated storage areas are used (e.g., cool rooms, refrigerators, etc.).
72. Documentation is available.
73. Food is covered where practicable while cooling down.
74. There are adequate controls to prevent the likelihood of cooked and ready-to-eat foods becoming contaminated by raw, unprocessed food.
75. Cooling down times and temperatures are satisfactory and monitored by staff.
76. Necessary steps are taken to prevent contamination during the cooling down process.

Packing/Service

77. The product packing/distribution forms are up to date and all corrective actions have been completed.
78. Records of temperature monitoring for all refrigerated strong areas are used (e.g., cool rooms, refrigerators, etc.).
79. Serving times and temperatures are satisfactory and monitored by staff.
80. Staff are aware of the risk of contamination.

Item	1, 2, 3, 4	Corrective Action
81. Staff are following good hygiene practices.		
82. Pest control measures are adequate and effective.		

Transportation

83. All foods are stored in suitable containers to maintain temperature control during transit.
84. The temperature of all food items is checked before distribution.
85. All food items are covered in such a way as to eliminate contamination.
86. All the hazards during transportation have been identified.

Complaints

87. A system to record nonconformances exists.
88. Customer complaints are recorded.

Product and Process Changes

89. Product specifications for all high-risk foods are available.
90. Changes to equipment have been approved and incorporated into the food safety program (e.g., cleaning, maintenance, calibration programs).
91. Changes to processes have been approved and incorporated into the food safety program.
92. Training records have been updated in accordance with system requirements.

Food Safety Program Management System

93. An internal food safety audit has been conducted.
94. The food safety program is up-to-date.
95. Amendments to the food safety program have been documented.
96. The flow chart is available.
97. The hazard analysis has been done.
98. The food safety plan is available.

REFERENCES

ANZFA (Australia New Zealand Food Authority) Food Safety: An audit system—An information paper outlining an audit system developed for the purpose of auditing food safety programs. www.foodstandards.gov.au/publications/documents/FS_Audit_Report_final%20edit0702.pdf.
GFSI. (October 2013). GFSI Guidance Document—Part IV: 165, 6th edn./Version 6.3.
ISO 22000: 2005, Food safety management systems: Requirements for any organisation in the food chain.
ISO 9001: 2008, Quality management systems: Requirements.
ISO/TS 22003: 2007, Food safety management systems: Requirements for bodies providing audit and certification of food safety management systems.
ISO/TS 22004: 2005, Food safety management systems: Guidance on the application of ISO 22000:2005.
Knaflewska, J. and Pospiech, E. (2007). Quality assurance systems in food industry and health security of food. *Acta Sci. Pol. Technol. Aliment* 6(2): 75–85.

7 Strategies for Achieving Food Safety

7.1 CONSUMER FIRST: RIGHTS AND LEGISLATIONS

The consumer is always at the center stage whenever the issue of food safety is handled. The safety, health, and welfare of the consumer are paramount whenever any regulations or legislations are enacted. The safety and wholesomeness of food has always been important for humankind. Given the intimate relationship between human health and food, the resolve to keep it protected has given rise to a situation that ensures that food has never been safer than it is today. Yet, food safety is an increasingly important global issue. This is not solely due to a rise in the number of reported food safety events, but it is also due to increased consumer awareness, increased globalization, and the complexity of the food chain. Trade of food and agricultural products around the world continues to increase, with worldwide food and agricultural exports more than doubling from $400 billion in 2000 to $900 billion in 2007. This growth in international trade has meant that food safety hazards that may have previously been confined to a relatively small area can now disseminate with ease across countries and continents. Accordingly, reacting to food safety events in isolation and after their occurrence is inadequate. Anticipation, prevention, and timely action should be the principal means to counter food safety threats (FAO, 2010). Consumers have many considerations, like quality and taste, but the ultimate concern still remains the safety of the food. Because consumers have different and sometimes conflicting interests in relation to food, there is a need to balance those conflicting interests. The focus is on three main themes that need to be addressed and balanced to get the best overall outcomes for consumers (FSA, 2015):

1. The right to be protected from unacceptable levels of risk
2. The right to make choices knowing the facts
3. The right to the best food future possible

From a national perspective, it is imperative that governments create an environment that facilitates the safety of their citizens. Governments have a responsibility and duty to protect their citizens and ensure that they are able to get safe food and stay healthy. In furtherance to this responsibility, every government has enacted legislation and regulations in food safety. These serve four main purposes:

1. To protect human health
2. To ensure that food placed in the market is safe
3. To safeguard consumers against deception
4. To ensure the public receives accurate information

7.2   CHALLENGES IN FOOD SAFETY

Challenges in food safety can be distinguished at three levels:

1. *Global/regional*: Strategies can be formulated around issues with the Technical Barriers to Trade Agreement (TBT) and the Agreement on the Application of Sanitary and Phytosanitary Measures (SPS); insufficient border inspections; the high cost of microbiological and chemical analysis; lack of accredited laboratories; and too-stringent food regulations.
2. *National*: A national food control system/plan; food laws, regulations, and legislation; the high cost of microbiological and chemical analysis; lack of accredited laboratories; untrained staff; and incompetent food inspectors doing their work without adding value to the industry.
3. *Facility/factory*: Lack of awareness of food safety and quality issues, and lack of incentives for operators in the value chain to invest in good practices and food safety management systems.

The key challenges in food safety facing any country/regional body or authority (EFSA, 2008) can be summarized as follows:

1. Globalization increases the likelihood of new or reemerging risks to the food supply.
2. There is a need to deal with innovative technologies, evolving risk assessment practices, and new science.
3. Sustainability and climate change will emphasize the importance of an integrated approach to risk assessment.
4. Societal changes associated with sociodemographic structure, diet, and consumer behavior will have an impact.
5. Changes in policies and the regulatory framework will have implications for workload and priorities.

Since innovations are happening, new technologies are evolving, and consumer preferences and demands are changing at every level, governments and regulatory agencies are trying to keep pace and are coming out with new solutions and regulations. The ultimate aim is that food and feed should be safe and wholesome. Community legislation comprises a set of rules to ensure that this objective is attained. These rules extend to the production and placing on the market of both feed and food (EU, 2004).

7.3   PURPOSE AND GOALS OF FORMULATING STRATEGIES

Strategies are developed in recognition of our primary responsibility and accountability to the consumer to protect public health and the other interests of consumers in relation to food. These strategies mainly relate to official controls delivery to determine and drive improvements and bring greater consistency in the approach to dealing with food risks across the food chain. The challenges of consistency and transparency are compounded by the complexity of the food chain and food trade. Keeping in view the levels and types of challenges being faced, the food control system and strategies are changing in such a way that they can address these issues, so these strategies and solutions are dynamic and evolving. Normally, the food safety strategy of any country or region will be dependent on the following:

1. Food laws, acts, and standards
2. Existing food control systems
3. Systems for control and prevention of foodborne diseases
4. Presence and effectiveness of surveillance systems
5. Appropriate food safety policies/plans of action
6. Technical capacity and financial resources
7. Alignment with international standards and alert systems

The Codex Alimentarius Commission (CAC, 2014) realized that, with increased globalization, the Commission must also be capable of responding in a timely manner to emerging food safety issues and other factors that may impact on food safety and fair practices in the food trade, such as the effects of shifting populations, climate change, and relevant consumer concerns. Food standards, guidelines, and recommendations established by the CAC are recognized as reference points for food under the relevant World Trade Organization (WTO) agreements. To advance the mandate of the CAC during the period 2014–2019, the Commission came out with the Strategic Plan 2014–2019. This plan presents the vision, goals, and objectives for the Commission and is supported by a more detailed work plan that includes activities, milestones, and measurable indicators to track progress toward accomplishment of the goals. The four major goals of the Strategic Plan 2014–2019 are as follows:

1. Establish international food standards that address current and emerging food issues
2. Ensure the application of risk analysis principles in the development of CAC standards
3. Facilitate the effective participation of all CAC members
4. Implement effective and efficient work management systems and practices

Many countries, in furtherance to the decision taken at the CAC meeting, came out with separate strategic plans specific to them. Keeping in view the previously mentioned four goals, the Strategic Plan for National Codex Committee (NCC)-India 2015–2019 has been formulated by the Food Safety and Standards Authority of India (FSSAI, 2015), which is coterminus with the CAC Strategic Plan 2014–2019. The purpose of the Strategic Plan 2014–2019 is mainly to meet the mandate of the CAC during the period 2014–2019. The Strategic Plan for NCC-India has been formulated consistent with the goals and the terms of reference finalized by the CAC. Unfortunately, the plan does not contain something specific to India, especially as the food safety landscape of India is not the same as that of developed western countries. There should have been an effort to address the issue of traditional and street food in a comprehensive way, as most of the population in India uses these kinds of foods (WHO, 2011). Although the Czech Republic has more similarity to Indian food safety landscape in activities, capabilities, and foods, still an effort has been made to prepare a strategic paper that suggests deviations as per the requirements of the country. The Czech Republic (Czech, 2010), based on the CAC Strategic Plan 2009–2013, realized that the main task shall be to review the existing system for ensuring food safety, primarily in the light of the effectiveness of its functioning. Against this background it is necessary to carry out a consistent evaluation of the interministerial cooperation standard upon which the system is built. Attention shall be paid to the assurance of risk assessment, cooperation with the European Food Safety Authority, and nutrition. The Strategic Plan for Food Safety by the Czech Republic identified four areas in which action needs to be taken (CAC, 2007):

1. Science-based health risk assessment
2. Risk management
3. Development of communication and education
4. Strategies for achieving food safety

7.4   STRATEGIES FOR ACHIEVING FOOD SAFETY

7.4.1   NATIONAL FOOD CONTROL SYSTEM AND DEVELOPING A NATIONAL FOOD CONTROL STRATEGY

When seeking to establish, update, strengthen, or otherwise revise food control systems, national authorities must take into consideration a number of principles and values that underpin food control activities, including the following:

- Maximizing risk reduction by applying the principle of prevention as fully as possible throughout the food chain
- Addressing the farm-to-table continuum
- Establishing emergency procedures for dealing with particular hazards (e.g., recall of products)
- Developing science-based food control strategies
- Establishing priorities based on risk analysis and efficacy in risk management
- Establishing holistic, integrated initiatives that target risks and impact on economic well-being
- Recognizing that food control is a widely shared responsibility that requires positive interaction between all stakeholders

The attainment of food control system objectives requires knowledge of the current situation and the development of a national food control strategy. Programs to achieve these objectives tend to be country specific. Like socioeconomic considerations, they are also influenced by current or emerging food safety and quality issues. Such programs also need to consider international perceptions of food risks, international standards, and any international commitments in the food protection area.

Therefore, when establishing a food control system it is necessary to systematically examine all factors that may impinge upon the objectives and performance of the system, and develop a national strategy.

7.4.1.1   Collection of Information

The collection of information is achieved through the collection and collation of relevant data in the form of a country profile (see Annexure 8). This data underpins strategy development, with stakeholders reaching consensus on objectives, priorities, policies, roles of different ministries/agencies, industry responsibilities, and time-frames for implementation. In particular, major problems associated with the control and prevention of foodborne diseases are identified so that effective strategies for the resolution of these problems can be implemented.

The profile should permit a review of health and socioeconomic issues impacting on foodborne hazards, consumer concerns, and the growth of industry and trade, as well as identification of the functions of all sectors that are directly and indirectly involved in ensuring food safety and quality and consumer protection. The collection of epidemiological data on foodborne illness is an indispensable component of a country profile and should be done whenever possible.

7.4.1.2   Development of Strategy

The preparation of a national food control strategy enables the country to develop an integrated, coherent, effective, and dynamic food control system, and to determine priorities that ensure consumer protection and promote the country's economic development. Such a strategy should provide better coherence in situations where there are several food control agencies involved with no existing national policy or overall coordinating mechanism. In such cases, a strategy prevents confusion, duplication of effort, inefficiencies in performance, and wastage of resources.

Devising strategies for food control with clearly defined objectives is not simple, and the identification of priorities for public investment in food control can be a challenging task. The strategy should be based on multisectoral inputs and focus on the need for food security and consumer protection from unsafe adulterated or misbranded food. At the same time it should take into consideration the economic interests of the country in regard to export/import trade, the development of the food industry, and the interests of farmers and food producers. Strategies should use a risk-based approach to determine priorities for action. Areas for voluntary compliance and mandatory action should be clearly identified and timeframes determined. The need for human resource development and strengthening of infrastructure such as laboratories should also be considered.

Certain types of food control interventions require large fixed capital investments in equipment and human resources. While it is easier to justify these costs for larger enterprises, imposing such costs on smaller firms who may coexist with larger enterprises may not be appropriate. Therefore, the gradual phasing in of such interventions is desirable. For example, countries may allow small enterprises longer periods of time to introduce a Hazard Analysis and Critical Control Point (HACCP) system.

The strategy will be influenced by the country's stage of development, the size of its economy, and the level of sophistication of its food industry. The final strategy should include

- A national strategy for food control with defined objectives, a plan of action for its implementation, and milestones
- Development of appropriate food legislation, or revision of the existing legislation to achieve the objectives defined by the national strategy
- Development or revision of food regulations, standards, and codes of practice, as well as harmonization of these with international requirements
- A program for strengthening food surveillance and control systems
- Promotion of systems for improving food safety and quality along the food chain (i.e., introduction of HACCP-based food control programs)
- Development and organization of training programs for food handlers and processors, food inspectors, and analysts
- Enhanced inputs into research, foodborne disease surveillance, and data collection, as well as creating increased scientific capacity within the system
- Promotion of consumer education and other community outreach initiatives

7.4.2 STRENGTHENING ORGANIZATIONAL STRUCTURES FOR NATIONAL FOOD CONTROL SYSTEMS

Given the wide scope of food control systems, there are at least three types of organizational arrangement that may be appropriate at the national level:

1. *Multiple agency system*: A system based on multiple agencies responsible for food control
2. *Single agency system*: A system based on a single, unified agency responsible for food control
3. *Integrated system*: A system based on a national integrated approach

7.4.2.1 Multiple Agency System

While food safety is the foremost objective, food control systems also have an important economic objective of creating and maintaining sustainable food production and processing systems. In this context, food control systems play a significant role in the following:

- Ensuring fair practices in trade
- Developing the food sector on a professional and scientific basis
- Preventing avoidable losses and conserving natural resources
- Promoting the country's export trade

The systems that deal specifically with these objectives can be sectoral (i.e., based upon the need for development of a particular sector such as fisheries, meat and meat products, fruit and vegetables, milk and milk products). These systems can be mandatory or voluntary and put into effect either through a general food law or a sectoral regulation. Examples include

- An export inspection law that identifies foods to be covered for mandatory export inspection prior to export, or offers facilities for voluntary inspection and certification for exporters
- Specific commodity inspection regulations, such as for fish and fish products, meat and meat products, or fruit and vegetable products, which are implemented by different agencies or ministries given this mandate under relevant law(s)
- Regulated systems for the grading and marking of fresh agricultural produce that goes directly for sale to the consumer or as raw material for industry (these are mostly confined to quality characteristics so that the producer gets a fair return for his produce and the buyer is not cheated)

Where sectoral initiatives have resulted in the establishment of separate food control activities, the outcome has been the creation of multiple agencies with responsibilities for food control. Typically, under such arrangements the food control responsibilities are shared between government ministries such as health, agriculture, commerce, environment, trade and industry, and tourism, and the roles and responsibilities of each of these agencies are specified but quite different. This sometimes leads to problems such as duplication of regulatory activity, increased bureaucracy, fragmentation, and a lack of coordination between the different bodies involved in food policy, monitoring, and control of food safety. For example, the regulation and surveillance of meat and meat products may be separate from food control undertaken by the Ministry of Health. Meat inspection is often done by the Ministry of Agriculture or primary industry personnel who undertake all veterinary activities, and the data generated may not be linked to public health and food safety monitoring programs.

Food control systems may also be fragmented between national, state, and local bodies, and the thoroughness of implementation depends upon the capacity and the efficiency of the agency responsible at each level. Thus consumers may not receive the same level of protection throughout the country and it may become difficult to properly evaluate the effectiveness of interventions by national, state, or local authorities.

7.4.2.2 Single Agency System

The consolidation of all responsibility for protecting public health and food safety into a single food control agency with clearly defined terms of reference has considerable merit. It acknowledges the high priority that governments place on food safety initiatives and a commitment to reducing the risk of foodborne disease. The benefits that result from a single agency approach to food control include (WHO, 2013)

- Uniform application of protection measures
- Ability to act quickly to protect consumers
- Improved cost efficiency and more effective use of resources and expertise

- Harmonization of food standards
- Capacity to quickly respond to emerging challenges and the demands of the domestic and international marketplace
- Provision of more streamlined and efficient services, benefiting industry and promoting trade

While a national strategy helps to influence both the legislation and the organizational structure for enforcement, it is not possible to recommend a single organizational structure that will universally meet the requirements and resources of every country's socioeconomic and political environment. The decision has to be country specific and all stakeholders should have the opportunity to provide inputs into the development process. Unfortunately, there are often few opportunities for countries to build a new food control system based on a single agency.

7.4.2.3 Integrated System

Integrated food control systems warrant consideration where there is desire and determination to achieve effective collaboration and coordination between agencies across the farm-to-table continuum. Typically, the organization of an integrated food control system would have several levels of operation:

1. *Level 1*: Formulation of policy, risk assessment, and management, and development of standards and regulations
2. *Level 2*: Coordination of food control activity, monitoring, and auditing
3. *Level 3*: Inspection and enforcement
4. *Level 4*: Education and training

In reviewing and revising their food control systems, governments may wish to consider a model that calls for the establishment of an autonomous national food agency that is responsible for activities at Levels 1 and 2, with existing multisectoral agencies retaining responsibility for Level 3 and 4 activities. The advantages of such a system include

- Provision of coherence in the national food control system
- Political acceptability as it does not disturb the day-to-day inspection and enforcement roles of other agencies
- Promotion of uniform application of control measures across the whole food chain throughout the country
- Separate risk assessment and risk management functions, resulting in objective consumer protection measures with resultant confidence among domestic consumers and credibility with foreign buyers
- Better capacity to deal with the international dimensions of food control such as participation in the work of the CAC, follow-up on the SPS/TBT Agreements, and so on
- Encouragement of transparency in decision-making processes and accountability in implementation
- Greater cost-effectiveness in the long term

Responding to these benefits, several countries have established or are in the process of creating such a policy-making and coordinating mechanism at the national level. Case studies demonstrating national food control systems in selected countries are presented in Annexure 9.

By placing management of the food supply chain under a competent, autonomous agency, it is possible to fundamentally change the way food control is managed. The role of such an agency is to establish national food control goals and put into effect the strategic and operational activities necessary to achieve those goals. Other functions of such a body at the national level may include

- Revising and updating the national food control strategy as needed
- Advising relevant ministerial officials on policy matters, including determination of priorities and use of resources
- Drafting regulations, standards, and codes of practice and promoting their implementation
- Coordinating the activity of the various inspection agencies and monitoring performance
- Developing consumer education and community outreach initiatives and promoting their implementation
- Supporting research and development
- Establishing quality assurance schemes for industry and supporting their implementation

An integrated national food control agency should address the entire food chain from farm to table, and should have the mandate to move resources to high-priority areas and to address important sources of risk. The establishment of such an agency should not involve day-to-day food inspection responsibilities; these should continue to lie with existing agencies at the national, state/provincial, and local levels. The agency should also consider the role of private analytical, inspection, and certification services particularly for export trade.

7.4.3 COMPLIANCE AND ENFORCEMENT STRATEGY

Principles and Approaches of Compliance and Enforcement Strategy are as described in the following.

7.4.3.1 Principles

The principles behind this compliance and enforcement strategy are

- To target interventions in areas where there is highest risk
- To give greater recognition to businesses' own means of securing compliance
- To increase the transparency of a business' food safety and hygiene standards
- To use wider incentives and penalties that drive compliance (recognizing the different drivers in different food sectors and/or businesses)

- To put more emphasis on tackling persistent noncompliance with swift action on serious noncompliance
- To·have a consistent, risk-based application of controls throughout the food chain and an increased focus on their outcomes

7.4.3.2   Approach

The approach of this compliance and enforcement strategy will (EU, 2005)

- Improve the approach to assessing and profiling risk across the food chain to target delivery of official controls
- Work internationally to deliver flexible, risk-based official controls
- Make use of the current flexibilities in the system of official controls for targeting enforcement activities
- Use accredited third-party assurance to inform the type and frequency of official interventions
- Assess how inspections undertaken by independently accredited, private organizations could be used in the food hygiene ratings scheme
- Support the continued take-up of the Primary Authority Scheme for food hygiene and standards
- Continue to promote the provision of information to consumers about food safety standards in businesses and monitor its impact on business compliance
- Increase understanding of what interventions work in incentivizing the right behaviors and tackling underlying behaviors and attitudes
- Put in place steps to ensure that noncompliance costs more for businesses than compliance
- Escalate quickly the enforcement activity on high-risk and/or persistent noncompliance, and make sure there are deterrents in place to influence food business perceptions around risk and the consequences of being caught
- Identify and prioritize criteria to assess risk-based planning and delivery of official controls

7.4.4   FOOD STANDARDS AGENCY STRATEGIC PLAN 2015–2020 (UK)

The Food Standards Agency (FSA) Strategic Plan 2015–2020 reinvigorates the pledge to put consumers first in everything the FSA does, so that food is safe and consumers can make informed choices about what they eat, now and in the future.

This strategy recognizes that there are growing challenges around food safety, affordability, security, and sustainability, and makes clear the FSA's purpose and responsibilities, and the roles and responsibilities of others, in meeting these.

The supporting FSA Strategic Plan 2015–2020 sets out the proposed approaches and outlines a number of key initiatives to take forward in Years 1 and 2, to make sure the FSA delivers on the priorities in the strategy and achieves its strategic outcomes to ensure that consumers are consistently protected, informed, and empowered.

These approaches include

- Using science, evidence, and information both to tackle the challenges of today and to identify and contribute to addressing emerging risks for the future
- Using legislative and nonlegislative tools highly effectively to protect consumer interests and deliver consumer benefits, influencing business behavior in the interests of consumers
- Being genuinely open and engaging, finding ways to empower consumers both through policy making and delivery, and in their relationship with the food industry

7.4.5 EMPRES FOOD SAFETY STRATEGIC PLAN

The EMPRES (Emergency Prevention System for Food Safety) Food Safety Strategic Plan is a holistic and multidisciplinary program that aims to prevent and deal with food safety emergencies at a global level by partnering with international, regional, and national agencies, as well as Food and Agriculture Organization (FAO) decentralized offices. Its broad approach is to work with existing initiatives to prevent, mitigate, and manage food safety threats. EMPRES has defined eight major elements of its strategic plan to achieve the program's aim. These elements have been grouped under the following three pillars of EMPRES food safety:

1. Early warning:
   a. Provide early warning: Engage with INFOSAN to provide early warning of food safety threats
   b. Conduct horizon scanning: Anticipate food safety threats through food safety analytical intelligence of low-key signals and indicators
2. Emergency prevention:
   a. Prevent escalation of imminent threats: Short-term, rapid addressing of imminent food safety threats to prevent the threat from occurring, escalating, or recurring
   b. Prioritize food safety threats
   c. Fill knowledge gaps
   d. Formulate and prepare prevention projects
   e. Provide tools, advice, and activities for preparedness
3. Rapid response:
   a. Conduct rapid response
   b. Provide timely responses to identified food safety emergencies (all activities are conducted in the broader scope of FAO normative food safety activities [FAO, 2010]):
      i. Capacity building
      ii. Scientific advice

The seven underlying principles of food safety are the basis for the formulation of good strategies, and understanding these helps to achieve food safety. These principles also illustrate the responsibilities and roles played within the network, and form the pillars upon which the food safety structure rests (Germany, 2008). The main principles are

1. The food chain principle (i.e., food safety involves a complete food chain)
2. The producers' responsibility principle (i.e., responsibility for providing safe food rests with food business operators (FBOs))
3. The traceability principle (i.e., all foods and their components should have a traceability trail)
4. Independent scientific risk assessment
5. Separation of risk assessment and risk management
6. The precautionary principle (refers to specific situations where there are reasonable grounds for concern that an unacceptable level of risk to health exists and the available supporting information and data are not sufficiently complete to enable a comprehensive risk assessment to be made).
7. Transparent risk communication

From a national perspective, it is imperative that governments initiate the conduct of a needs assessment for food safety capacity building that can be implemented at the systems, organizational, and individual levels. Generally, the assessment process requires the following steps:

1. Review and analyze the current capacity or situation
2. Define the desired future of the food safety systems
3. Identify gaps in abilities or areas for improvement
4. Prioritize those needs
5. Identify options to address the needs, including assistance from external support
6. Undertake monitoring and evaluation

Food safety programs must ultimately be able to prevent exposure to unacceptable levels of foodborne hazards along the entire food chain. They should aim to bring scientific objectivity and balance to food safety initiatives. Innovative approaches must be adopted to solve problems, and these initiatives must be in place to advocate and assist in the development of a risk-based, sustainable, and integrated food safety system. The program should also enable the government to effectively and promptly assess, communicate, and manage foodborne risks/crises, which requires concerted efforts by all relevant stakeholders. All food safety systems have their own constraints, but what must be done is to find ways to work effectively within these constraints and move aggressively to remove those constraints that limit a government's ability to protect the public's health. When it comes to food safety, there is not one single solution; instead there should be a series of sensible approaches formulated to address the different situations in different countries. It is also important that these efforts be undertaken in a concerted manner to improve the food safety system.

7.5 SUGGESTIONS FOR IMPROVEMENTS IN PLANNING DELIVERY OF SAFE FOOD

Establishing food safety is of paramount importance for the food industry, thus requiring and ensuring that FBOs adhere to food safety norms and standards. The current structure and performance of market surveillance, especially in developing nations that are not in a position to invest in laboratories and capacity building, requires a lot to be done. The private sector food safety and quality assurance systems—more so in countries that have more contributions from medium and small enterprises—are largely characterized by informal production and trade have the responsibility to improve. This requires educating all those concerned from the farm level up to the delivery of finished goods at the customer level. Whenever such food safety strategies are to be finalized, the agri-business reality in developing countries and the issues of the food security cannot be separated out or ignored. Globally, we need to work toward building efficient and effective cross-border food supply chains. There are constraints that prevent regulatory compliances. Apart from the requirements of time and money, there are also a number of more complex issues, including information bottlenecks, absence of regular updating of standards, lack of trust in food safety legislation, trust deficit with enforcement officers, lack of motivation in dealing with food safety legislation, and lack of knowledge and understanding.

Many efforts, including provision of food safety training to FBOs, capacity building in regulatory affairs, motivation of FBOs in the implementation of good practices of food safety standards, information sharing between different agencies, and empowerment of consumers can help to establish a better food safety control system. The impacts of these efforts may vary from one economy to the next due to the different characteristics of financial and managerial capabilities, so a unique approach or strategy may be required in different economies. As has been explained previously, every country and every region has its unique strengths and weaknesses. There are commonalities in challenges, purposes, and goals between countries, but still a common strategy may not be applicable to all. There can be some similarities in approach and principles, but each country has to have specific plans and solutions. This is not only confined to countries or regions; strategies may differ between areas within the country and also may vary from facility to facility as well (WHO, 2014).

1. *Global/regional*: Strategies can be formulated around issues with the Technical Barriers to Trade Agreement (TBT) and the Agreement on the Application of Sanitary and Phytosanitary Measures (SPS); insufficient border inspections; the high cost of microbiological and chemical analysis; lack of accredited laboratories; and too-stringent food regulations.
2. *National*:
    a. Upgrade the policy frameworks, especially the policies toward food safety, security, and public sector organizational structures
    b. Create and upgrade networks at national levels as well as for representation in international forums and standard setting organizations
    c. Upgrade the institutional capacities, referring to the development of the organizational capacities of public and private entities that have a stake in food safety and quality assurance
    d. Have coordination between food processing associations and the government and formulate a national food control system as per the needs and requirements of the country

3. *Facility/factory*:
   a. Human resource development is crucial for the success of a food safety plan, so
      i. Sensitize everyone in the facility toward food safety
      ii. Involve external consultants/accredited agencies to train staff
      iii. Have standard protocol for training and updating the skills of staff
      iv. Monitor the effectiveness of training programs
      v. Invite all concerned staff into production meetings and discuss issues so that they all have information about other sections as well
   b. Support cooperation and networking through
      i. Linkages between food operators at certain stages of the food supply chain (horizontal) and along the food supply chain (vertical)
      ii. Cordial relations between food supply chain operators and service providers and between private and public stakeholders

Empowerment of FBOs requires comprehensive approaches to resolving issues such as limited access to information, education, training, technology, capital, and market access. Networking and public–private partnerships are good platforms to enhance FBO empowerment via education, training, and appropriate communication to improve awareness of food safety.

However, without going into the detail about the differences and challenges that each area, region, or country has to face, there are a few suggestions that may be relevant at one stage or another as per the requirements, and thus can be adopted accordingly. The suggestions are as follows:

1. The large majority of countries in the region must try to give higher priority to capacity building to respond to the unacceptable burden of illnesses caused by the consumption of unsafe food. The capacities in terms of manpower and laboratories are abysmally low or almost nonexistent in some areas or countries.
2. Capacity-building programs need to identify and take into account the needs of targeted FBOs. A capacity-building program should stress the importance of changing the behavior of FBOs' human resources toward development of food safety culture.
3. A well-coordinated, multisectoral approach to food safety risk analysis at all levels is the key for the preparation and success of any strategy.
4. All the resources available at private, public, university, or institute levels must be utilized by regulatory authorities.
5. There should be coordination and information sharing between established surveillance systems.
6. Analysis of regulation impact on FBOs is an exemplary approach to finding the appropriate regulatory avenue to encourage FBOs to comply with food safety regulations and any amendments and updates to them.
7. It is essential to include consumers in food safety education to help in reporting food safety incidence.

8. The FAO/World Health Organization (WHO) need to support countries, especially in Asia and Africa, in areas of assessment on food control capacity, capacity building, surveillance coordination, and food safety strategy for the region.
9. More coordination and information sharing on food trade, imports/exports, and risk-based inspections is required.
10. Communication strategies to FBOs should take advantage of the various media that are applicable to/suitable for the literacy level of the FBO, and to have better reach these should be in the local language. The content of such communication should emphasize the importance of food safety to promoting business development. Ideally, successful communication strategy and capacity building may be scaled up and replicated in other locations.
11. Strengthen Good Agricultural Practices (GAP) through the development of schemes and certification system initiatives to address food safety challenges.

APPENDIX 7.A

The development of a national food control strategy calls for the collection, collation, and evaluation of the following type of information.

STATUS OF FOOD AND AGRICULTURE SECTOR

- Data and information on: primary food and agriculture production; food processing industry (*i.e.* types and number of establishments, processing capacity, value of production etc); food distribution and marketing.
- Information on formal (organized) and informal (cottage or household units, street-foods) industry.
- Potential for industry development.
- Food chain, and identification of key intermediaries who may influence quality and safety of foods.
- Market infrastructure including assets and deficiencies.
- Safety and Quality management programmes including level of HACCP implementation in the industry.
- Food consumption data. Information on consumers will include data on energy/protein intake, percentage of the population dependent upon subsistence economy, per capita income, etc.
- Cultural, anthropological, and sociological data is also important, including information on food habits and food preferences.

FOOD SECURITY, FOOD IMPORTS, AND NUTRITIONAL OBJECTIVES

- Food demand for nutritional requirements; postharvest food losses; type and quantities of food imports.

CONSUMER CONCERNS OR DEMAND

- Consumer demand on safety, quality and information (labelling) issues.

Food Exports

- Quantity and value of food exports and potential for growth in export trade.
- Data on detentions or rejections of food exported.
- Information on number and types of complaints from buyers and remedial action.
- Identification of foods with potential for export and target countries for export.

Epidemiological Information

- Information on prevalence and incidence of foodborne disease; procedures used for investigating and notifying foodborne diseases; information on food incriminated; suitability of collected data for risk assessment purpose.

Food Contaminants Data

- Information on prevalence and level of contamination of food; monitoring programmes for biological and chemical contamination of food; suitability of collected data for risk assessment purpose.

Human Resources and Training Requirements

- Information on the number and qualification of personnel involved in food control, that is, staff engaged in inspection, analysis and epidemiological services; information regarding on-going training, and educational activities; projections on future staffing and training needs.

Extension and Advisory Services

- Information on the existing extension and advisory services for the food sector as provided by the government, industry, trade associations, non-governmental organizations, and educational institutions; train-the-trainer type of activities; training needs analysis.

Public Education and Participation

- Consumer education initiatives in food hygiene; potential for increased involvement and interaction between government, consumer associations, non-governmental organizations, and educational institutions in risk communication activities; risk communication to prevent foodborne diseases and possible improvements.

Government Organization of Food Control Systems

- Listing of government departments and authorities concerned with food safety and food control activities.
- Description of the food control system and an overview of the resources, responsibilities, functions, and coordination between the entities; methods of determining priorities for action; options for raising resources.

FOOD LEGISLATION

- Current food legislative arrangements, including regulations, standards, and codes of practice.
- Information on authorities empowered to prepare regulations and standards, and how they coordinate their activities and consult with industry and consumer organizations.
- Capacity to carry out risk assessment.

FOOD CONTROL INFRASTRUCTURE AND RESOURCES

- Organization of inspection, surveillance, and enforcement activities (national, provincial, and local levels).
- Number and qualifications of inspection personnel.
- Resources within inspection agency, and assessment of strengths and weaknesses. Analytical support facilities (number of laboratories, facilities and equipment, monitoring programmes, etc).
- Codes of hygienic practice.
- Licensing arrangements for food premises.

Source: Reproduced from Food and Agriculture Organization of the United Nations and World Health Organization, *Assuring Food Safety and Quality: Guidelines for Strengthening National Food Control Systems*, FAO Food & Nutrition Paper, 76, FAO, Rome, Italy, 2003, www.fao.org/docrep/006/y8705e/y8705e00.htm. With permission.

APPENDIX 7.B

An improved understanding of the risk-based approach and growing awareness about the impact of food safety on public health and national economies has led many countries to make significant changes to their food control systems, in recent years. This response to the need for consumer protection against newly identified food hazards, coupled with the need for efficient use of public resources have practically forced national authorities in many industrialized countries to give priority to this task.

The current food scenario, particularly the work of the Codex Alimentarius Commission and the comparatively recent WTO Agreements on SPS and TBT, have reinforced the need for appropriate scientific inputs in food control decision-making processes and accelerated the review and reorganization of systems in many countries. While this is an on-going exercise and several changes are still in the offing in many countries, it is useful to study a few newer models or approaches that are emerging in the important area of strengthening food control infrastructure.

All countries that have revised or updated their food control systems expect benefits in terms of increased efficiency, greater ability to provide farm-to-table oversight of food safety, and enhanced international market access. Apart from enhanced objectivity in protection measures and consolidation of their activities, there is a move by governments to shift the responsibility for ensuring food safety to the food industry, with governments assuming an audit or oversight role.

CASE STUDY: CANADA

Background

Food safety in Canada is a shared responsibility between the Federal Government (Health Canada and the Canadian Food Inspection Agency (CFIA)), provincial/ territorial governments, the food industry and consumers.

The Canadian food safety system has been developed in a way that enables it to keep abreast of rapid changes in the nature of food, increased globalization of the food trade, and changing public expectations of food safety. Three fundamental principles underpin the food safety system:

(a) the health of the population must remain paramount;
(b) policy decisions must be grounded on scientific evidence; and
(c) all sectors and jurisdictions must collaborate to protect consumers.

Regulatory Framework

The *Food and Drugs Act* is the principal federal legislation covering food safety, and it prohibits the manufacture or sale of all dangerous or adulterated food products anywhere in Canada. The Act derives its authority from criminal law, and is supplemented by regulations designed to ensure food safety and nutritional quality.

Other federal trade and commerce legislation references the *Food and Drugs Act* and imposes additional requirements *e.g. Canada Agricultural Products Act, Meat Inspection Act, Fish Inspection Act, Feeds Act* and *Pest Control Products Act*, etc.

At the local level, provinces and territories are responsible for public health, including food surveillance, investigations and compliance. Therefore, provinces and territories also enact legislation to control foods produced and sold within their own jurisdictions. These laws are complementary to federal statutes. As legislative power cannot be delegated from one level of government to another, governments collaborate in areas of shared jurisdiction, such as food inspection, and establish partnerships to ensure effective and efficient program delivery. Provinces and territories legislation also authorize municipalities to enact by-laws on food inspection.

Health Canada

Development of Standards and Policies

Health Canada sets standards and policies governing the safety and nutritional quality of all food sold in Canada. Government has the primary responsibility for identifying health risks associated with the food supply, assessing the severity and probability of harm or damage, and developing national strategies to manage the risks. Canada has adopted a risk analysis process that provides a common, consistent, comprehensive, and scientifically sound mechanism to identify, assess, and manage potential risks to public health. Accordingly, all food-policy decisions are made in a transparent and rational manner.

Health Canada also carries out foodborne disease surveillance activities providing a system for early detection and a basis for evaluating control strategies.

To ensure the federal system is one with checks and balances, the Minister of Health has responsibility for assessing the effectiveness of the Agency's activities related to food safety.

Canadian Food Inspection Agency: Enforcement and Compliance

The Canadian Food Inspection Agency (CFIA) is responsible for enforcing federal food safety policies and standards.

Mission and Objectives

In order to fulfil its mission of *Safe Food, Market Access and Consumer Protection*, the CFIA has adopted the following objectives:

(a) to contribute to a safe food supply and accurate product information;
(b) to contribute to the continuing health of animals and plants for protection of the resource base; and
(c) to facilitate trade in food, animals, plants and their products.

Activities

The CFIA was created in 1997 and delivers federal inspection services to food safety as well as plant protection and animal health. The agency operates under the authority of 13 federal acts and 34 sets of regulations, and meets it responsibilities through 14 distinct programs.

CFIA is responsible for all federally mandated food inspection, compliance and quarantine services. Prior to 1997 these activities were undertaken by Agriculture and Agri-Food Canada, Health Canada, Industry Canada and Fisheries and Oceans Canada. CFIA develops and manages inspection, enforcement, compliance, and control programs and sets service standards. It also negotiates partnerships with other levels of government and non-government organizations (NGOs), as well as industry and trading partners, with respect to inspection and compliance programs; and supplies laboratory support for inspection, compliance, and quarantine activities. CFIA also issues emergency food recalls, and conducts inspections, monitoring, and compliance activities along the food continuum. The Agency is supported by a national network of service laboratories.

The CFIA is responsible for the administration and enforcement of the following acts: Agriculture and Agri-Food Administrative Monetary Penalties Act, Canada Agricultural Products Act, Canadian Food Inspection Agency Act, Feeds Act, Fertilizers Act, Fish Inspection Act, Health of Animals Act, Meat Inspection Act, Plant Breeders' Rights Act, Plant Protection Act, and Seeds Act. The Agency is also responsible for enforcement of the Consumer Packaging and Labelling Act and the Food and Drugs Act as they relate to food, and the administration of the provisions of the Food and Drugs Act as they relate to food, except those provisions that relate to public health, safety or nutrition which remain under the responsibility of the Minister of Health.

Food Safety Partnerships

Coordination of the activities in Canada's food control system is exercised through a number of committees and established Memoranda of Understanding. For example, both Health Canada and the CFIA are involved in international activities related to food safety. Coordination of these activities is achieved through a Health Canada/CFIA Committee on International Food Safety. The Canadian Food Inspection System (CFIS) is a federal-provincial-territorial initiative to facilitate national harmonization, streamline the inspection process, and reduce regulatory pressures on industry. This initiative is managed by the Canadian Food Inspection System Implementation Group (CFSIG) which has membership representing the federal government (Health Canada and CFIA) as well as the governments of the provinces and territories (WHO, 2002).

There are several mechanisms for facilitating cooperation among governments, industry, academia, consumers, and NGOs in Canada. Through an Integrated Inspection Systems approach, the CFIA works with food manufacturers and importers to develop and maintain a Hazard Analysis Critical Control Point (HACCP) system. The goal of the CFI's compliance and enforcement activities is to move away from dependence on government inspections to increased use of government audits of industry activities. The audits are based on risk, supported by strong compliance and enforcement tools. The degree of ongoing government oversight and intervention depends on each company's history of compliance and the risk associated with its product.

The CFIA also facilitates the development of safety programs along the entire food continuum through programs such as the Canadian On-Farm Food Safety Program.

The Federal Provincial Territorial Committee on Food Safety Policy, under the leadership of Health Canada, develops, coordinates and provides leadership on food safety policies and standards, educational programs and the exchange of food safety information on issues of regional, national, and international importance.

The Canadian Food Inspection System (CFI) is a federal-provincial-territorial initiative to facilitate national harmonization, streamline the inspection process, and reduce regulatory pressures on industry. Harmonization with international standards is an objective of all CFIS initiatives.

The Food-borne Illness Outbreak Response Protocol is a partnership among provincial and territorial governments, Health Canada, and CFIA that describes an integrated response to national and regional foodborne disease outbreaks, causing high levels of severe morbidity or mortality. The Protocol ensures that all responsible agencies are notified promptly and work collaboratively to mitigate and contain risks.

Risk-Based Approach

Health Canada has adopted a decision-making framework that provides a consistent, and comprehensive means of identifying, assessing, and managing risk. Similarly, CFIA has also adopted a risk-based approach to enforcement, compliance, and control processes. The concept of precaution is an intrinsic part of Health Canada and CFIA's risk analysis process. Uncertainties in scientific data are carefully considered in assessing the level of risk to which the public may be exposed and in the selection of an appropriate risk management strategy.

Risk management is accomplished through the establishment and enforcement of legislative and regulatory requirements, as well as the application of nonregulatory options such as guidelines, advice and education, and promotion of voluntary compliance by industry.

A number of factors are considered when selecting an appropriate risk management response, including legislative authority, international trade obligations, national policies, and feasibility, as well as socio-economic factors such as culture, consumer concerns and demographics.

Further Developing the Food Safety Framework

The Government of Canada is reviewing how to optimize operational efficiency and ensure stakeholder participation in food safety.

The factors that prompted Canada to review and restructure its food inspection system are not unique to Canada. The need to make more efficient use of limited public resources while ensuring the consumer is adequately protected are challenges faced by both developed and developing countries.

CASE STUDY: IRELAND

Background

Aside from public health considerations, the importance of food production to the Irish economy necessitates independent and verifiable assurances as to the quality and purity of its food products. As a result, the Irish Government initiated a review of their food safety systems in 1996.

The outcome of the review was a recommendation to establish the Food Safety Authority of Ireland (FSAI) as a statutory, independent, and science-based body, overseeing all functions relating to the food safety regulation of the food industry. On 1 January 1999, the Food Safety Authority of Ireland was formally established under the *Food Safety Authority of Ireland Act, 1998*. The Act

(a) Established the Authority as an independent body accountable to the Minister for Health and Children

(b) Transferred all responsibility for ensuring compliance with food safety legislation to the FSAI

(c) Conferred powers on the Authority (which included those powers available under existing food safety legislation, as well as additional new enforcement powers)

(d) Provided that the existing food control enforcement arrangements at local and national level would remain in place, but would be carried out under 'contract' to the FSAI by various public bodies involved in food safety services delivery

(e) Provided mechanisms for the FSAI to keep food safety service delivery under review and to report to the Minister for Health and Children in relation to such matters, in particular on the scope for better coordination and delivery of the food inspection services

Mission

The Food Safety Authority of Ireland's mission is to protect consumers' health by ensuring that food consumed, distributed, marketed, or produced in Ireland meets the highest standards of food safety and hygiene.

Structure

The Authority is a statutory, independent, and science-based body, governed by a Board of ten members appointed by the Minister for Health and Children. A Consultative Council gathers the views of stakeholders involved in the production and consumption of safe food. A Scientific Committee prepares scientific advice on food safety issues. Decisions on food safety and hygiene take account of the latest and best scientific advice and information available from independent experts.

The FSAI is led by a Chief Executive who supervises a multidisciplinary team including many specialists, for example, public health practitioners, veterinarians, food scientists, environmental health specialists, microbiologists, public relations personnel, etc.

Food Safety Authority of Ireland Organizational Chart

Operations

The principal function of FSAI is to take all reasonable steps to ensure food produced, distributed, or marketed in Ireland meets the highest standards of food safety and hygiene reasonably available and to ensure that food complies with legal requirements, or where appropriate with recognized codes of good practice.

The FSAI operates the national food safety compliance programme by means of service contracts with agencies currently involved in the enforcement of food legislation. This includes the Department of Agriculture, Food and Rural Development, the Department of the Marine and Natural Resources, Department of Environment and Local Government, as well as regionally based Health Boards and Local Authorities.

FSAI is also responsible for promoting communication, education, and information on food safety matters (risk communication). This includes establishing and managing public relations and promotional activity, and developing and implementing policy on communication, education, and information for consumers, industry, and enforcement officers.

Food Safety Environment

Responsibility for creating food safety policy (risk management) lies with a number of Ministers of the Government, with the Minister for Health and Children having the coordinating role. Scientific advice upon which policy decisions are taken (risk assessment) is obtained from the Scientific Committee of the Food Safety Authority of Ireland. Food safety services (risk management) are delivered through a number of different Government Departments and agencies at national, regional, and local level. The Food Safety Authority is responsible for ensuring the coordinated, effective, and seamless delivery of food safety services by those agencies.

The enforcement of food safety legislation relating to on farm activities is not within the scope of the FSAI. The Department of Agriculture, Food and Rural Development and Department of the Marine and Natural Resources enforce such legislation.

Source: Reproduced from Food and Agriculture Organization of the United Nations and World Health Organization, *Assuring Food Safety and Quality: Guidelines for Strengthening National Food Control Systems*, FAO Food & Nutrition Paper, 76, FAO, Rome, Italy, 2003, www.fao.org/docrep/006/y8705e/y8705e00.htm. With permission.

REFERENCES

CAC. (2007). Codex Alimentarius Commission Strategic Plan 2008–2013. Rome, Italy: Food and Agriculture Organization of the United Nations.

CAC. (2014). Codex Alimentarius Commission Strategic Plan 2014–2019. Rome, Italy: Food and Agriculture Organization of the United Nations. ftp://ftp.fao.org/codex/Publications/StrategicFrame/Strategic_plan_2014_2019_EN.pdf, Accessed on February 15, 2016.

Czech Republic. (2010). Food Safety and Nutrition Strategy for 2010–2013. Prague, Czech Republic: Ministry of Agriculture of the Czech Republic. www.mze.cz, e-mail: info@mze.cz, Accessed on November 13, 2015.

EFSA. (2008). Strategic Plan of the European Food Safety Authority for 2009–2013. Adopted in Parma, Italy, December 18, 2008.

EU. (2004). Regulation (EC) No. 882/2004 of the European Parliament and of the Council of 29 April 2004 on official controls performed to ensure the verification of compliance with feed and food law, animal health and animal welfare rules. *Official Journal of the European Union* L191: 1–52.

EU. (2005). Regulation (EC) No. 183/2005 of the European Parliament and of the Council laying down requirements for food hygiene. *Official Journal of the European Union* L35: 1–22.

FAO. Assuring Food Safety and Quality: Guidelines for Strengthening National Food Control Systems. Joint FAO/WHO Publication, FAO Food & Nutrition Paper 76. www.fao.org/3/a-y8705e.pdf, Accessed on March 14, 2017. www.fao.org/docrep/006/y8705eoj.htm.

FAO. (2010). EMPRES Food Safety Emergency Prevention System for Food Safety Strategic Plan. Rome, Italy: EMPRES Food Safety Food and Agriculture Organization of the United Nations. http://www.fao.org/docrep/012/i1646e/i1646e.pdf, Accessed on November 13, 2015.

FAO/WHO. (2003). *Assuring Food Safety and Quality: Guidelines for Strengthening National Food Control Systems*. FAO Food & Nutrition Paper, 76. Rome, Italy: FAO, www.fao.org/docrep/006/y8705e/y8705e00.htm, Accessed on March 14, 2017.

FSA. (2012). Food Standards Agency, Compliance and Enforcement Strategy 2010–15: Aims, principles and approach. http://www.food.gov.uk/sites/default/files/multimedia/pdfs/enforcement/compliance.pdf, Accessed on February 15, 2016.

FSA. (2015). Food Standards Agency Strategic Plan 2015–20, FOOD we can TRUST. http://www.food.gov.uk/sites/default/files/Strategy%20FINAL.pdf, Accessed on February 15, 2016.

FSSAI. (2015). Strategic Plan of the National Codex Committee (NCC). http://www.fssai.gov.in/Portals/0/Pdf/NATIONAL_CODEX_COMMITTEE_STRATEGIC_PLAN.pdf, Accessed on February 15, 2016.

Germany. (2008). Food Safety Strategies, Bundesministerium für Ernährung, Landwirtschaft und Verbraucherschutz (BMELV) 11055 Berlin (Federal Ministry of Food, Agriculture and Consumer Protection). www.bmelv.bund.de, Accessed on November 14, 2015.

WHO. (2002). Global strategy for food safety: Safer food for better health. Geneva, Switzerland: World Health Organization. http://www.who.int/foodsafety/en, Accessed on November 13, 2015.

WHO. (2011). Draft report of the Regional Consultation on safe street foods. New Delhi, India: World Health Organization Regional Office for South-East Asia.

WHO. (2013). Strategic Plan for Food Safety Including Foodborne Zoonoses 2013–2022. Advancing Food Safety Initiatives. WHO Press. http://www.searo.who.int/entity/food-safety/global-strategies.pdf, Accessed on February 15, 2016. ISBN: 9789241506281.

WHO. (2014). Regional Food Safety Strategy, 2013–2017. New Delhi, India: Regional Office for South-East Asia.

Index

Milton Keynes UK
Ingram Content Group UK Ltd.
UKHW040104071024
449327UK00019B/791